# PRAISE
# *REBUGGING TH*

This is a lovely little book that could and should have a big impact. The decline of insect life in the UK and globally is one of the biggest concerns of our biodiversity crisis. We often feel so helpless about nature loss, so it's hugely inspiring to find out that there is something we can actually do about it. Let's all get rebugging right away!

**Hugh Fearnley-Whittingstall**,
multi-award-winning writer and broadcaster

A bold and educational call to action and call to arms in one of the most crucial challenges facing society – halting the dreadful destruction of the amazingly little animals we call invertebrates or bugs. Time to get rebugging!

**Matt Shardlow**, author and chief executive
of Buglife – The Invertebrate Conservation Trust

Everyone should read Vicki's delightful bug book! Like me, she's a Londoner, but unlike me, she's realised that her lifelong fascination for nature in general and insects in particular can be explored in an urban setting. Her passion for bugs is palpable and wonderfully illuminated through individual bug stories.

**Patrick Holden**, CBE organic farmer,
and founding director of the Sustainable Food Trust

What a fantastic, timely and important book! For too long, our society has taken bugs for granted when in reality they represent the very foundations of our food system, our economy, our civilisation. With her well-researched but personable and highly readable writing style, Vicki Hird offers an engaging and hopeful narrative about what we can and must do to make insects matter. In doing so, she doesn't just stick with the easy stuff like what needs to happen in your garden or local park. She also tackles the need for system level change; in agriculture, in politics, in the economy, in culture – all while gaining fascinating insights from the remarkable world of insects.

**Craig Bennett**, chief executive of the Wildlife Trusts

*Rebugging the Planet* is a joyous and impassioned song to the insect life on which we all depend. Brimming with wisdom but accessibly written, it is a call to arms to avert Insectaggedon. Without bugs, we're in deep trouble!

**Guy Shrubsole**, environmental campaigner and author of *Who Owns England?*

Packed with eye-opening facts and leaving not a stone unturned in her efforts to understand and explain the causes of their decline, Vicki inspires each and every one of us to re-evaluate our relationship with these magnificent minibeasts. This book provides us with the tools and advice we need to 'rebug' our gardens, our lives and our world.

**Brigit Strawbridge Howard**, author of *Dancing with Bees*

Hird's joy in bug life is infectious and her knowledge encyclopaedic. I defy even the most bug-phobic reader not to finish Hird's book without, if not sharing her love of them, at least joining in her admiration. If you've ever asked what bugs have done for us, read this book – and then join the movement to protect them!

**Caroline Lucas**, Green Party MP

In 1987 E.O. Wilson told us that 'bugs' were the little things that run the world. We didn't listen and instead have forced millions of species of these essential creatures to the brink of extinction. Just in time, Vicki Hird tells us how and why we need to change our cultural relationship with 'bugs' and reverse these disastrous declines. Despite the serious nature of this subject matter, *Rebugging the Planet* is a light-hearted and delightful read.

**Douglas W. Tallamy**, author of *Nature's Best Hope*

This book is a delightful exploration into the world of 'bugs' – replete with creative use of words like 'rebug', 'rewild', 'insectageddon', 'invertosphere', 'entomophage' (the practice of eating bugs) and 'fog basking'. Full of colourful stories about specific novel species like the cockchafer, the hummingbird hawkmoth and money spiders, it is also a call to action to do everything we can to stop the modern assault on bugs.

**Stephanie Seneff**, author of *Toxic Legacy*

# Rebugging the Planet

*The Remarkable Things that Insects (and Other Invertebrates) Do – And Why We Need to Love Them More*

VICKI HIRD

FOREWORD BY GILLIAN BURKE

Chelsea Green Publishing
White River Junction, Vermont
London, UK

All illustrations courtesy of Ned Page.

Cover photographs by (*clockwise from top-left*) iStock/imv, Chris Clor, iStock/arlindo71, Luxy Images, iStock/aluxum, iStock/marcouliana, Athiwat Poolsawad/EyeEm, iStock/Antagain, iStock/kjohansen.



Project Manager: Alexander Bullett
Commissioning Editor: Jonathan Rae
Developmental Editor: Muna Reyal
Copy Editor: Susan Pegg
Proofreader: Diane Durrett
Indexer: Linda Hallinger
Designer: Melissa Jacobson
Page Layout: Abrah Griggs

Printed in the United States of America.
First printing September 2021.
10 9 8 7 6 5 4 3 2 1 21 22 23 24 25

**Library of Congress Cataloging-in-Publication Data**
Names: Hird, Vicki, author.
Title: Rebugging the planet : the remarkable things that insects (and other invertebrates) do – and why we need to love them more / Vicki Hird.
Description: First. | White River Junction, Vermont : Chelsea Green Publishing, [2021] | Includes bibliographical references and index.
Identifiers: LCCN 2021029967 (print) | LCCN 2021029968 (ebook) | ISBN 9781645020189 (paperback) | ISBN 9781645020196 (ebook)
Subjects: LCSH: Insects.
Classification: LCC QL463 .H57 2021 (print) | LCC QL463 (ebook) | DDC 595.7—dc23
LC record available at https://lccn.loc.gov/2021029967
LC ebook record available at https://lccn.loc.gov/2021029968

Chelsea Green Publishing
85 North Main Street, Suite 120
White River Junction, Vermont USA

Somerset House
London, UK

www.chelseagreen.com

*To my sons, Tom and Ollie, my husband, Tim,*
*and the many fantastic people working hard to*
*protect the bugs, their habitats and their protectors*
*across the globe, on farms and down the street.*

# CONTENTS

# FOREWORD

They knew I was watching. My head hovered just above the nest entrance for a close-up view. If I held my breath, the workers seemed happy to carry on with their chores, using their dainty mouthparts, like tiny cement mixers, to knead and mould their mud tunnels. Then, inevitably, I would have to breathe. As I did, the soldiers, with their oversized heads, would detect my presence and pour out to make it abundantly clear I was getting too close. Not a problem, though. All I had to do was back off and 'termite watch' could resume. As long as I kept a respectful distance, the termites would settle down again and carry on with their daily tasks of building, feeding and protecting the colony.

Even as a child, I was struck by the different shapes and sizes of workers in the colony, and by the division of labour between the workers and soldier termites – and all this in just *one* colony! More than anything, though, it was my first clue that us humans aren't all we're cracked up to be.

This was me, growing up in Kenya in the early 1980s, barefoot and device-free, where playing was watching termites, dodging *siafu*, or driver, ants, catching shiny-black ground beetles and making offerings of sweet fresh grass with all the care a five-year-old could muster. I had none of the knowledge to inform me that my little carnivorous pet beetles were never going to eat grass and, with a horrifying 'attrition rate', my attention moved on to *jongoo*, or millipedes, with their determined single-track progress, made all the more mesmerising by their crimson legs that seemed to move in a perpetual wave. Fortunately for the millipedes, I had let go of my insect-farming dreams, but I couldn't resist a gentle prod so I could watch them coil up in a perfect, protective spiral.

I don't know where this fascination with invertebrates came from. I had no access to field guides, and there certainly wasn't anything special about me. The only reasonable conclusion is they were fascinating simply because they were *there*. Sadly, this is no longer true for many children, particularly in urban areas and in the Global North. And it's not just a loss for happy childhood memories and free play: this is an existential threat to all of humanity.

Globally and year-on-year, we've lost 10 per cent of all insects since 1970.[1] At this rate, we will have lost 80 per cent of insects by 2050.[2] Zoom in on some of the finer detail and the data shows that in parts of Europe we've already overshot that mark. Studies in Germany and the Netherlands have found a midsummer decline of flying insect biomass of 82 per cent over 27 years of record keeping.[3] The curve continues in an ever-steepening, stomach-turning nose-dive with complete ecosystem collapse within our lifetimes in sight.

Like a determined marching band striding against this bleakest of backdrops, *Rebugging the Planet* serves up optimism and realism with a whopping big dose of practical advice as to what we can all do to reverse the trends. From activities for children to grassroots activism to lobbying your MP and big corporations, all the bases are covered. Vicki Hird has mapped out a landscape in which all the touch points into the Lilliputian invertebrate world are clearly posted.

We all need a little 'rebugging' in our lives. That's because words and language matter. When 'rewilding' was coined, that single word blazed a trail for a whole new way of thinking, where we followed nature's lead in restoring whole systems and finding balance. Keystone species like wolves, bears and beavers have been the superstar flag-bearers of the rewilding movement; however, and it is no exaggeration to say, all the big, sexy megafauna and flora stand on the shoulders not of giants but on the unsung heroes of the invertebrate world. *Rebugging* creates a banner for a movement that recognises and celebrates the myriad tiny and unseen creatures that outweigh, underpin and hold the whole beautiful tapestry of life together. *Rebugging* holds space for the urgent need to provide not just a literary home but a literal home for all little spineless wonders of the world.

What about my beloved termites? A quick internet search will tell you that termites are among the most destructive insects in the world. What it doesn't tell you about is the magical moment that the winged males and virgin queens take flight, before a long-awaited downpour, with glittering, glass-like wings that crinkle and sparkle in flight like nature's own confetti. This is how I remember them, as they amped up an already charged atmosphere with a palpable blend of excitement and relief for everyone, because after a long and hot dry season the 'rainflies' heralded the coming of the rains. Crops would grow and people would eat.

GILLIAN BURKE, co-presenter of BBC's
*Springwatch* and vice president of BugLife

# Introduction

*If you want to live and thrive, let the spider run alive.*

An old proverb I learned as a child

I was never going to get the pony I wanted, so I settled for an ant farm at an early age.

I have no idea where this interest in bugs and nature came from. Maybe it was my grandparents, who had a beautiful garden and looked out for the birds. But, despite being featured in our local paper at the age of eleven as a birdwatcher with a pair of binoculars around my neck, I was never as keen on birds as bugs. My eyes were downwards focused. Ants featured frequently in my garden wanderings, to the frustration of my mother, who poured boiling water on the nests near the house. But the ants kept coming and their social behaviour was deeply fascinating. Where were they all going and why did they carry the dead bodies of fellow ants around? I collected them from the garden and kept them in an old ice cream tub in my room to more easily observe them. However, I was untrained in the art of ant care and they failed to thrive, died or made their escape from the inhospitable world of a girl's bedroom.

My first literary introductions to bugs were the Collins insect guides and *The Country Diary of an Edwardian Lady* by Edith Holden, which I was given for my birthday. This was a posthumous publication of Holden's observations and poetry, alongside charming drawings of birds, plants and insects. While I had no gift for drawing, I found the book captivating and my interest with invertebrates has never left. With inspiring teachers I was encouraged to study biology. One teacher secured me a post-exam summer job at a local research institute, which was a dreamy alternative to shelf stacking. I spent many hours with one of the world's leading bee experts, counting bees in and out of hives, with or without pollen sacks. We were testing which pheromones (the chemical signals bees use to communicate) encouraged foraging and which triggered fight or flight behaviours. At that point, my insect love was probably fixed forever – as anyone who has studied bees cannot fail to fall in love with them.

In later years, I investigated many pest species including rats, aphids, leaf miner flies and cockroaches, the last of which I developed a deep respect for. These unfairly maligned insects are remarkably sleek, and fast, too, as I found when I tried to catch escapees in the laboratory. They are highly adaptable, able to live in a huge variety of habitats and feed on many different foods. Some even produce milk for their babies. I was investigating ways to control them but, really, it is us whom we need to control better.

I have spent around thirty years as an environmental campaigner, researcher and lobbyist, and through all that time, invertebrates have been such a strong motivator for carrying on when despair could easily have set in. Having children reinforced for me the need to protect the planet they would inherit. The additional joy in seeing your child's complete fascination for a worm was a major bonus. We bred stick insects as pets for my young boys, which I know they will always remember. And my insect passion remains very much alive – I had a giraffe-necked weevil (unique to Madagascar) tattooed on my shoulder for my fiftieth birthday.

But the past decades of overuse of the world's resources have been hugely damaging, despite the work of campaigners, scientists and communities to protect the environment and the natural beauty of this world. There is growing evidence of a major crisis in invertebrate populations and it's clear that we can't carry on with business as usual.

Less scientifically, but potentially driving new public interest, those of a certain age (including myself) have noticed the strange absence of bug-splattered windscreens. When I was young and went on family trips through England, the windscreen and headlights of our car would be thick with dead bodies when we arrived at our destination. We also see far fewer butterflies or wasps around when we picnic or stroll in the countryside – the iconic stars of a huge cast of species that underpin life on earth have seemingly vanished.

This may seem overdramatic, and invertebrates as a whole are unlikely to go extinct, yet many studies at a national, and even global, scale are showing crashes in both the number and diversity of insects and other bugs. One recent study in 2019 drew from 73 reports of insect declines from around the world, echoing many other studies showing a disturbing trend. Their review suggested that over 40 per cent of insect species are in decline and so at risk of extinction over the next decades, more than twice that of vertebrate species.[1] There were strong critiques of the study methods, but previous analyses have shown similar declines but received less attention. We also don't really know what we are losing. In addition to the one million identified types of insect, there may be over four million yet undiscovered species. And that is just the insects. Millions of other invertebrates, on land and sea, are also undiscovered. We have not yet catalogued far more species than those we have recorded, and they may be lost through deforestation and other actions before we get chance to do so.

So most global analyses are beginning to indicate that we are seeing a major loss of numbers and diversity of species worldwide as well as locally – and even global extinctions.[2] In the UK alone, twenty-three bee and wasp species have become extinct since 1850, while the number of pesticide applications, a key factor in wildlife harm, has almost doubled in the twenty-five years from 1995.[3] According to Buglife, an organisation formed in 2000 to champion the invertebrate cause, in the UK: 'butterflies, moths, bees, wasps, and dung beetles are amongst the most at risk, along with freshwater insects such as stoneflies, caddisflies and mayflies.' It is a depleted world, which we are creating. And, as I finish this book in 2020, the world has been turned upside down by the Covid-19 pandemic. Scientists are warning that this pandemic has revealed how far we have disrupted the natural systems through forest destruction, industrial-scale farming and the pushing of small-scale farmers further out to the margins.

We should learn the lessons from this, and also from invertebrates, how to fit into and live with nature, rather that assume we are above it and can fix any threats through science and technology.

So, what do I mean by 'rebugging'? My crucial proposition is that we can all rewild by rebugging, and that there is far more to rebugging than site-based actions – we need to rebug our lives, too. Rewilding is mainly defined as the reintroduction of almost-natural systems, and often missing species, into areas and then leaving nature largely to take care of itself. It has become an extremely popular and often controversial issue, given the huge pressure on land use, but there are inspiring examples of rewilding which I explore more in chapter 3.

But for me rebugging means this and more. We also need to join with others and act as citizens, to make the bigger policy changes. It matters how we live, how we buy stuff and how we engage in society.

This book also aims to gladden hearts with great tales and learnings of the invertebrate world, bring awareness of their demise and, finally, give readers the tools to act. It does not pull punches when it comes to the difficult, political, social

## What do I mean by 'bugs' and 'rebugging'?

Let me take a moment to explain. The word 'bug' is often used to refer to tiny creatures that crawl along, such as insects and even small animals that are not insects: spiders, millipedes, worms and water-dwelling creatures. Scientists use the word bug in a more specific way to mean insects that have mouthparts adapted for piercing and sucking; these are known as 'true bugs' and include aphids, cicadas, spittle bugs (they sit in that spit you find on plant stems) and shield bugs.

For this book, I use a broader biological definition, which may seem more familiar: bugs are small creatures that do not have a vertebral or spinal column – called 'invertebrates' – and which are in the arthropod (insects, arachnids, crustaceans and myriapods) and annelid (earthworms and leeches) families. I occasionally stray into other taxonomic groups, such as slugs, for reasons you will discover.

and economic issues. But if it means you notice more bugs and grubs in your life, and if it inspires you to do something, it has done a decent job.

This book is about helping everyone to do a citizen rebug. We can do this.

## Imagine a world without bugs

Over the past few years, the global media have been reporting on a so-called 'Insectageddon' and what a bugless world could mean. They cite the growing body of evidence that the invertebrates, and particularly the insects, are in big trouble.

This can make for scary reading, which can make some people feel powerless. But it is having an impact on research budgets and government action which were much needed, and at least the media are also starting to explain to the public, more helpfully, why this decline will be a problem: telling the story of bugs in our lives and describing what it will mean if we lose them. That we will lose many of the foods that we take for granted, including coffee, chocolate and fruits. That these invertebrates are the butterflies we love and that they provide food for the birds we also love. If these go, we have lost not only the means to feed us but much of the beauty and reasons to enjoy life itself...

So, a chilling story, but is it a sensationalised media response to the scientific evidence? A little bit, yes. If you took some of the reportage at face value, and you have the means, you may be considering stockpiling food, breeding bees and building a fortress against the coming crisis. We are not quite at that stage yet, but it could start to get close if we don't act now. For someone like me who has long yearned for greater interest in and support for invertebrates, this attention is welcome. So, I'd like to try and show what Insectageddon could be like – especially as we are already seeing some signs of collapse.

*'If we die, we're taking you with us'*

A great image that has been doing the rounds is a picture of a bee saying 'If we die, we're taking you with us.' Invertebrates are the glue that binds the plants, microbes, fungi and animals to each other on this small planet, and we can quite safely say that we would not last long without invertebrates.

Loosing even a small amount, a tiny percentage, of bug life could be catastrophic locally. Bugs sit at the bottom of the food web; if they disappear, so will the species that feed on them. We would lose many bigger animals such as birds, bats, some mammals, fish, reptiles and amphibians that we've come to love, and which mean so much for our identity and culture. Whole ecosystems and even landscapes will change in a cascade of impacts we can't even imagine.

Bugs are a vital part of the recycling of nutrients, without which we cannot survive. The soil in which we grow most of our food is created largely by the guts and jaws of worms, mites, springtails, termites, beetles and many more. They mash up the leaf litter and the dead bodies, so we don't have to, releasing some nutrients and making plant material more easily decomposed by fungi and microbes, which then releases more vital nutrients like sugars, nitrates and phosphates for plants to absorb and grow.

And, without many bugs, most plant pollination would be impossible save for some carried out by the wind and a few reptiles and mammals (if they don't need the bugs, too, that is). But these larger beasts won't fit into a buttercup or a bluebell. The intricate way in which plants and bugs have evolved together is extraordinary and largely irreplaceable. To be blunt, without pollinating bugs and other beasts (which, in turn, need bugs to survive), almost 90 per cent of our flowering plants would die off.[4] This would have a catastrophic impact on ecosystems worldwide, not to mention the food on our plates. The world would be drained of its colour.

*Robo-bees*

A third of the crops we eat – and I am not talking just the basics here like fruit and vegetables, but also those essentials, such as chocolate and coffee – need invertebrates for their pollination. Some human and machine pollination is used: in China, for instance, where wild bee colonies have disappeared in some areas, workers pollinate orchards using brushes. But they can only cover a tiny proportion of crops and it is expensive. If we had to perform all that pollen transfer ourselves or with machines it would take a vast army of workers or a whole new level of robot insects.

We may be able to engineer tiny robo-bees in their billions to try and do the job, but they will never be as good, or as cheap, self-replicating and non-polluting as the real bees and the flies and the moths. Such robots are actually already being produced in laboratories. A 'RoboBee' has been developed at Harvard University for artificial pollination, potential rescue services and, possibly, military surveillance, too. But as the expert bee academic Professor Dave Goulson points out:

> *Consider just the numbers; there are roughly 80 million honeybee hives in the world, each containing perhaps 40,000 bees through the spring and summer. That adds up to 3.2 trillion bees. They feed themselves for free, breed for free, and even give us honey as a bonus. What would the cost be of replacing them with robots?*[5]

Billions of pounds is the answer, yet we can get these services for free, or via relatively cheap honeybee colonies. The use of tiny robots could also create a pollution disaster and an additional vulnerability into our food system. And what would the animals that feed on the bees like birds and mammals then eat – the metal robots?

One of our sweetest gifts from the insect world, honey, would no longer be available in a world without bees. Synthetic substitutes can never taste the same, given honey's complex makeup, which includes vitamins, minerals, pollen, fragrance compounds, and even antibacterial and antifungal agents. Nor would we likely be able to grow enough sugar-forming plants to replace the honey because the soil would be so badly depleted without the critical and complex work of invertebrates.

Bees are only one part of the pollination picture. Our plates would be so much duller with a vastly reduced selection of plants pollinated by many invertebrates – no broccoli or sprouts, tomatoes or raspberries, to name only a few that would be gone. And the soil, which the bugs keep healthy and fertile with their industrious activities, would be so weakened that it is likely that many non-pollinated plants could not grow either.

If we lose hoverflies, ladybirds and wasps, which are fantastically effective predators of many real pests such as aphids, we could also see huge infestations and yields plummeting. More food disasters.

Maybe we will be able to grow some of our food and fibres in sterile systems or factories without needing soil. The nutritional implications of getting all our food this way is difficult to imagine. We are only now realising how vital the complex soil microbiota – the communities of protozoa, viruses, bacteria, fungi and bugs – are for providing the nutrients we need to ensure gut health. And how it's the multiple connections between these microbiota and the growing plants that are exchanging, delivering and mixing the huge range of nutrients, from sugars to trace minerals, in the system we rely on. Recreating this dynamic, complex web formed over many millennia in a lab may just prove impossible.

## A changed landscape

The world would also look so different without the bugs. The joy of a buzzing meadow would be a distant memory, seen only in films showing how the world around us used to have such vibrant colour, sounds and smells. The flowers and trees that rely on bugs would no longer light up our lives as we walk the streets or provide shade on a hot day. But we may also be walking in ever-growing mounds of our own waste as we would have none of the fantastic invertebrates that clean up our mess and digest our poo.

You may be surprised to know that we use bugs in sewage treatment plants to filter and break down matter, and help neutralise toxins. Without them, chemicals would be the primary way we'd clean water and I've no idea how we'd manage that without poisoning ourselves.

The intimacy with which our lives, even our skin, are entwined with invertebrates is too rarely considered. From the bed we sleep in and the clothes we wear, to the clean water we need to drink and wash with. We take it for granted. What would we wear if we had no invertebrates to ensure the grass grew for the sheep and cattle or to help the cotton crops thrive? With no wool, leather or cotton, we would have to resort entirely to artificial fibres. How long before the pollution from plastic microfibres irreversibly damages the rivers and seas?

And to complete this picture... what would we sit on? Have you sat on a wooden seat, used a wooden table or been in a house with wooden floorboards today? With no bugs to maintain the soil for healthy tree growth, there would be little new timber to produce furniture, buildings or even paper. Plastic, metal or concrete sofa, anyone? If you used any wood today, the tree it came from needed that army of bugs – some for nutrients, others for pollination and then more to spread its seeds, or as food for the birds that spread the seeds.

## Time to take action

It is a deliberately bleak picture I am painting. I do not think it will come to this. We have stepped back from the brink in the past. The book widely recognised as one of the most important environmental publications of the twentieth century, Rachel Carson's *Silent Spring*, was published in 1962. Carson, both a scientist and a renowned author, wrote carefully and passionately about the impact of indiscriminate use of insecticides – the chemicals being used to control crop and human pests. Back then, the world listened to her call to action. The evidence was strong and well presented, and she made careful proposals. Despite staunch industry resistance, those in power made some significant changes to legislation, created new

## What are the main causes of decline?

There is no simple or single reason for the alarming losses of invertebrates we have started to see across the globe. We are still unsure, for instance, what caused mass bee colony declines in North America, despite serious money being thrown at the problem. But from the evidence and long-term studies, the drivers fit into the following areas:

- industrial farming of crops and livestock, mining and urban development have destroyed the habitats invertebrates need for food, mating, egg laying and shelter, and have removed the natural corridors (or created barriers such as roads and developments), which prevent species from moving, colonising and mating
- air, soil and water pollution, mainly by synthetic pesticides and fertilisers
- biological factors, such as the introduction of new diseases or species which can predate or outcompete the domestic varieties
- climate change, including temperature shifts and extreme weather events
- light and noise pollution and even 5G and wireless signals that may be cooking the bees [6]
- microplastics filling the bugs' stomachs

agencies and eventually banned to use of the most harmful chemicals – the organochlorines – from most applications.

But, almost sixty years later, the alarm bells are ringing loudly again, everywhere, and the problems (and the humans and their consumption) have multiplied. Invertebrates have an intrinsic right to thrive, but regardless of whether you appreciate this, given how essential they are to us, we need to act.

## The great rebugging challenge

You are probably getting the picture. We need to rebug. Let me stop describing what a world without invertebrates would look like. I would far rather talk about how rebugging our lives could reboot our relationship with nature – and why it is such a brilliant idea.

This book is about how we can all rebug the planet and be part of the change we need to see. Rebugging will be like using a vacuum cleaner in reverse, undoing the sucking we have done on the land for too long. Scattering back those tiny vital bits onto the green and not-so-green carpet of the planet.

### *How this book can help*

I will show you how through rebugging every part of our lives – from small to large scale – we can help rewild our lives. And not just in nature, but also culturally and economically – and politically, if you are up for it. Because one stark observation I can draw from my work is that we need to change whole systems, not just bits of them, to ensure we can live in harmony with all the creatures on this planet.

We need to protect the entire habitat for invertebrates, not just the edges. Move to smart pest management using natural tools and knowledge to control pests, not just banning the worst chemicals while continuing to spray other toxic

**The only sea dwelling insect**

While many invertebrates fill the seas, from crabs to seas cucumbers, there is only one insect that calls the ocean its home. It has a wonderful common name: the sea strider. This carnivorous insect sprints on the water surface looking for prey that has fallen onto the water, such as zooplankton, fish eggs, larvae and dead jellyfish. In turn, it provides a source of food for sea birds and surface feeding fish.

pesticides around. Campaigns to protect a habitat for one rare species are valuable, but they are seriously not enough. Everything must shift. The rewilding movement feels like a strong expression of this need, but it must be about all the land, everywhere. Allowing nature to recover both itself and us in the round – and allowing us to reconfigure our relationship with it – to help save the life systems on which we depend.

The first two chapters of this book start with us. We have to rebug our attitudes and cherish the bugs, not only as fellow inhabitants of our planet, but by appreciating the work they do to make it comfortable for us. It is by re-evaluating our relationship to bugs, individually and societally, that we can begin to make the fundamental shifts in individual action, as well as policy-level changes to effectively rebug our world. We need to see bugs as citizens of our planet, with the same right to live and thrive as all humans and creatures.

The next chapters look at bugs within our ecosystem: how bugs are an intrinsic part of rewilding and must be part of our efforts to restore the natural world. I hope that chapter 4 will help you to do this at home and in your local environment, while the later chapters go big picture – how environmental changes have affected bug life and how we need to re-examine our food and shopping habits, and therefore our farm and agricultural practices.

The shocking Covid-19 pandemic was horrific and has rightly created some vital space for an urgent call to rethink our relationships: with the land, with the stuff we buy, with forests and with their inhabitants. Researchers have been pointing out for years that the continued destruction of forests, humans encroaching on more wild spaces with farms and infrastructure, and the impact on wild animal populations increases the risk of pandemics. This is now, rightly, coming under greater scrutiny.

In the end, though, rebugging has to happen at the highest level of society. We need political and economic structural changes, and chapter 7 examines the systemic challenges – and solutions – to rebugging. Issues of power and inequality may seem far removed from rewilding but they are critical. Finally, I end with a vision of what a rebugged world could look like, as well as ideas and names of organisations that will help you get involved at all levels of the great rebugging challenge.

I have woven rebugging tips throughout the book, some of which are simple and easy to do, while others will take a bit more effort. So, even if you have only a little time, there is still always something you can do. Small things you can do every day or every month, and some which can even save you time and money. However, if you have more time to do the things that need a bigger effort, you may find that you discover new communities and make new connections along the way. In the final chapters of the book, I have included ideas which involve a greater use of your time and resources and, if you want, your power as a citizen with influence. We all have some power.

And, above all, appreciate that although you are just a large, annoying, but occasionally useful, obstacle in the way of most invertebrates, you can still help them to help you.

— CHAPTER 1 —

# Rebugging our attitudes

Let's start with ourselves. We all need to love invertebrates more. We need to appreciate why they matter. Compared to the large furry mammals and the whales and dolphins, our interest in invertebrates – including beetles and ants, moths and worms – has been slow in coming. Even now, we have a limited understanding of invertebrates, their role in the ecosystems we depend on and what we may be losing as a consequence of their decline. Research on bugs is all too often about ways to control agricultural or domestic pests with chemicals and other means, rather than how to protect and nurture them for what they bring us. Too few baseline studies of populations exist globally to know what's there and what is disappearing. The business of researching insect diversity has been neglected for too long probably because people preferred the megafauna like tigers and pandas – important though they are. Thankfully, that is changing as public interest in invertebrates grows.

In 2020, a group of 30 international scientists expressed deep concern about global insect declines and urged action to not only to reverse this, but also to engage civil society in understanding the value of insects for our well-being.[1] They recognised that making insects popular was core to helping reverse the decline.

## Equal status for bugs

In a unique move, a municipality in Costa Rica's capital, San José, decided in 2014 to create a whole new relationship with its wildlife in order to protect and enhance its presence.[2] The canton of Curridabat is a dense urban area bearing little resemblance to the biodiverse wilderness and rich forests outside of the city, and which Costa Rica is famous for.

The then mayor of Curridabat, Edgar Mora Altamirano, decided to confer citizenship on all its pollinators, trees and native plants, and to work to ensure that they thrived by providing an urban haven for humans and wildlife. His administration also measured the diversity of plant, bird and invertebrate species to track progress as the detailed plans continue for a city connected with nature. In an interview with the *Guardian* newspaper, Edgar Mora explained how 'Pollinators were the key,' saying 'Pollinators are the consultants of the natural world, supreme reproducers and they don't charge for it. The plan to convert every street into a biocorridor and every neighbourhood into an ecosystem required a relationship with them.'[3] The team behind this project, called Ciudad Dulce, meaning 'Sweet City', have undertaken many activities, ranging from new urban pollinator-friendly garden design and innovative water and waste management initiatives to training local citizens to be Sweet City advocates. The project has won multiple awards for their exceptional efforts to link citizens with conservation and create truly living corridors for nature.[4]

Similar exciting urban initiatives are appearing elsewhere to welcome invertebrates. Monmouth in Wales has been named the UK's first 'Bee Town' with a 'nature isn't neat' programme. There are communities actively involved in keeping wildflowers in verges, parks and gardens, and avoiding harmful sprays and raising the profile of pollinators.[5] The US Bee

City affiliate scheme is also growing – cities pledge to provide pollinators with insecticide-free habitats and native plants.

Imagine if each city, town or borough did the same, and led its whole community towards understanding the joy and value of invertebrates. Such a wonderful example of urban regeneration needs replicating the world over.

## Love in the early years

One of my best memories of my youngest child as a toddler is his proud, excited face when he opened his pudgy little hand and showed me the soil-encrusted worm inside. He had just dug it up in the garden – he thought these wriggling wonders were amazing. He has retained a love of bugs throughout his life: filling his palms with crickets as we climbed mountains in Switzerland and, even as a jaded teen, I found him opening the back door with his elbow, hands cradling a spider that he had rescued from the bath. He is rebugging on a small scale.

Why do children love bugs? Do they recognise the huge value insects have for us as fellow dwellers on this small planet? Unlikely. I suspect it stems from pure curiosity in their small size and totally different appearance, combined with a lack of fear – valuable traits sadly lost to many of us as we mature.

My reaction to my son's discovery of dirt-covered worms was also important. We can instil love and fascination, or fear and disgust, pretty fast in impressionable young minds. I was thrilled and excited too. But then I am a bug obsessive. I

### Stick insects at home

Keeping stick insects, from the phasmids family, as pets can be highly rewarding. You can buy them from reputable breeders and keep them inside a large plastic tub with air holes – particularly useful if you don't have a garden. Stick insects are vegetarian and you can find their food, like bramble and ivy, in gardens and parks. They are great for children to observe insect behaviour and, as they can breed without mating (and males are rare), you may find yourself with a large number after a while, which you can then share with friends or safely destroy by freezing! I kept the common Indian stick insect, *Carausius morosus*, but there are many to choose from. Do get advice on care and handling from animal welfare organisations such as the Royal Society for the Prevention of Cruelty to Animals in the UK (www.rspca.org.uk).

hope, though, that I did not bore my son too much about the role of worms. I could have risked spoiling the pure, curious interest and fascination of a child in a tiny creature by lecturing him on the important tasks that bugs perform. Or be a doom-monger by lamenting about how they are threatened and how badly we have treated them. For young children, discovery, play and experience are enough. We must give them plenty of chances to be outside in nature and to experience bugs in their lives as much as possible.

Yet, as children get older, all too often that early curiosity is turned to fear for all the wrong reasons. Prevailing 'grown-up' attitudes start to dominate how children think of bugs, usually with negative associations, such as:

- wasps sting
- bees sting more
- spiders bite
- flies spread diseases
- slugs eat the flowers
- ants bite
- locusts destroy crops
- earwigs will climb into your ear

And so on. This conditioning may start quite early and eventually many children begin to get 'the fear'. They start to dislike bugs, assuming the bugs will hurt them or carry deadly diseases. They can become indifferent to what fascinated them. And there are also fewer bugs to see these days. A sad state, but reversible. Scientists have recognised the need for public appreciation of bugs as fundamental in driving societal changes and policy shifts.

What if adults, carers, everyone, understood the importance of bugs and, even if they did not find them fascinating, they understood their value? What if we all managed to put a different, positive message across? All the time. If everyone cannot love bugs for their beauty and extraordinary lives, perhaps they can appreciate the incredible gifts they provide us with, for example:

- wasps control pests and pollinate plants
- bees pollinate crops and make honey

- spiders catch flies
- flies clear waste and pollinate food plants
- slugs make soil
- ants aerate soil and clear away debris
- locusts are vital sources of protein and insects are eaten by two billion people globally
- earwigs pollinate fruit

And so on. Rebugging activities – like creating pollinator-friendly window boxes or putting on local bug treasure hunts – could help in this learning process, as well as provide greater interactions with nature and the joy that it can bring. It is also easier to observe these minibeasts in real life than the big beasts only seen at a zoo or on TV – bugs can be found all around us and creep into our everyday lives in other ways.

**Go on a bug treasure hunt**

Set up a search for bug-related treasures in your garden, park or anywhere with a bit of greenery. You can design a fun set of habitat and visual clues, and tasks to complete, such as making drawings or taking photos, so that children can learn about these creatures. You could find a leaf miner's wiggly track in a leaf, an ant busy tending aphids on a plant, a bulging gall (ball-like swelling) on a twig caused by a gall wasp, or a worm cast in among the grass.

## Bugs in puns, art and culture

Bugs have contributed to our language and culture in immeasurable ways. Where would we be without insect puns, idioms and poetry? We are metaphorical 'busy bees', but must not 'drone' on about it. Given how mechanical drones are now widely used for a growing number of tasks, it's odd to think the word drone is borrowed from the name for a male bee, who mates with the queen bee but has no other duties.

The great boxer Muhammad Ali said to win a fight he would, 'Float like a butterfly, sting like a bee,' which is the 'bee's knees' of a demonstration of insect similes. We talk increasingly of the 'hive mind' and like to surf the 'worldwide web', but do not like it when a 'bug' 'worms its way' into our IT systems.

We may hover round the biscuit tin 'like moths to a flame' and the person who treats you badly is a real 'louse'. The astonishing metamorphosis of insects from larval stage to adult is frequently used as metaphor, from literary pieces such as Kafka's *The Metamorphosis* to the description of a child's transformation to adult. Roman poet Ovid's great work *Metamorphoses*, written in around 8 AD, was an epic undertaking, comprising 15 books that chronicle the creation of the world to the death of Julius Caesar told through mythological stories. But there are also many examples of insect terms being used pejoratively in relation to people, such as the outdated, deplorable use of 'swarming' and 'cockroaches' to describe migration and refugees.

Words do matter and one of the greatest wordsmiths, Shakespeare, understood the invertebrate role well. He included many invertebrates in his plays, including beetles (*Measure for Measure*, act 3, scene1) carrion maggots (in *Hamlet*, act 2, scene 2), and silk moth cocoons (in *Othello*, act 3, scene 4) to fly metamorphosis (in *Love's Labour's Lost*, act 5, scene 2). And of course, we are all but food 'for worms'

(*Henry IV*, part 1: act 5, scene 4). When we rebug our world, maybe we can include more positive bug puns in our spoken and literary lives too.

## Up close and beautiful

Most invertebrates have an extraordinary beauty, from the tiniest rotifer swimming in water to the biggest ragworm burrowing in the mud. Their incredible forms and structures are a product of evolution with functional adaptations, but the sheer range of shapes, colours and textures are unmatched in the vertebrate or plant kingdoms. And this does not only apply to the stunning butterflies, bees and dragonflies, amazing though they are, but also to the iridescent wing case of the dung beetle and the incredible colours of the sea slug. Flies, too, can be extraordinarily beautiful up close.

As such, invertebrates present an irresistible, though probably rather skittish muse for any artist. What they create also provides inspiration, from elaborate spider webs and silk cocoons, to the intricacies of a mud-based termite mound or a paper-thin wasp nest. The cave paintings in the Cuevas de la Araña in Bicorp, Spain, date back around ten thousand years and depict bees flying around a person collecting honey. Beetle and butterfly pictures adorn the walls of ancient Egyptian tombs, and Leonardo da Vinci's sketch books have pages devoted to insect forms. Though he may have been seeking more scientific answers – such as learning how to fly from dragonflies – he still clearly recognised their beauty.

The fine line between science and art often blurs when it comes to insects, given their amazing structures and capacity for complex behaviours. The sensational *Micrographia*, published by Robert Hooke in 1665, revealed, among other highly meticulous observations, the detail of invertebrate morphology to an astounded public. Hooke used the first accurate microscopes to see and draw insect structures, mouthparts and skin in remarkable detail. The book was a bestseller with diarist Samuel Pepys declaring it, 'The most ingenious book that ever I read in my life.'

Before photography, artists were able to draw and paint invertebrates in their ecological settings – often beautiful but also a vital tool for scientists and the public to understand nature and ecology better. In the twentieth and twenty-first centuries, thousands of artists continue to make invertebrates their muse, and we can all even capture their beauty on our camera phones. British photographer Levon Biss takes it to another level entirely. He works for four weeks to produce an incredibly detailed image of insects using a very shallow depth of field to reveal the detail and complex beauty of a beetle wing case or a fly antenna. His 2017 book *Microsculpture: Portraits of Insects* is astounding.

## What we can learn from bugs?

Invertebrates are so numerous in number and variety, and inhabit almost all parts of the planet. They have lived and evolved here for over 650 million years, from the first marine invertebrate sponges. This compares to human life, which dates back only about two hundred thousand years. Over that amount of time, invertebrate adaptations have allowed them to live on most habitats on earth.

Evolution has not only delivered ideal designs of individual species, but has also optimised their interaction with each other and with other species. It has created a close relationship between their form and function and the biological, chemical and physical world around them, to allow for all needs. A worm is so perfectly adapted to life in the soil, with its smooth body able to burrow efficiently, with bristles on each body segment allowing it to push against the soil and so move quickly. A hummingbird hawkmoth's pipe-like tongue is precisely adapted to reach the parts of a flower other insects cannot reach. The adaptations involved are extraordinarily complex, but they can demonstrate to us smart ways to live more sustainably which are worth a brief look.

## What bugs can teach us

The truth is that bugs are exquisite in their evolutionary design and for hundreds of years humans have learned from bugs and applied this knowledge to human endeavours. Here are some examples to consider:

- The extraordinary tensile strength of spiders' thread made up of nanofibrils – comparable to that of high-grade alloy steel, but also able to stretch without breaking. Studies of its properties are helping us develop new 'nano' materials that work at extremely small scales.[6]
- The tough, protective shells that protect a beetle's wings, such as those of the wonderfully named 'diabolical ironclad beetle', have intricate layers to make them almost unbreakable and are a fantastic tool for developing high-strength but lightweight structures.[7]
- The ability of social insects like the ant and termite to organise themselves via complex

communication methods, by using chemicals or electricity as well as sound and visual cues. This is being used in learning how to communicate more efficiently. An 'ant colony optimization' (ACO) algorithm is inspired by real ant behaviour where they use chemical signals called pheromones to guide other ants. This algorithm is used to help develop computer learning. A bee version is based on how honeybees communicate when foraging.

- The ingenuity of termite nest construction and organisation is providing engineers with valuable design and construction insight, not least its sheer size (over two thousand times that of a termite), with nest walls that precisely regulate heat exchange.
- The organisational complexity of the colonies of social insects, such as ants and termites, is comparable in some ways to that of human society. Considerable research is looking at understanding swarm intelligence and self-organisation that we can learn from – including how we can better communicate, give feedback, and share labour and responsibilities. Some of the ideas resulting from this research have been successfully translated into useful tools in business, such as optimising workflows, allocating space in web servers and 'bucket brigade' worker ordering – a method

for transporting items where items are passed from one (fixed) person to the next.

- The smart engineering of hover, or flower, flies that have evolved ways to fly with incredible dexterity including forwards, backwards, sideways, up and down, and of course, they can hover. Leonardo da Vinci, the great renaissance artist and inventor, studied insects as well as birds to learn how to make flying machines – and the engineers of today are continuing those studies.[8]

  Flies are expert at flying and landing upside down – using visual cues such as looming objects on the eye, highly specific body manoeuvres and expert wing action – and all very swiftly. Researchers are aiming to use a machine-learning approach to teach the fly's movements to mechanical drones by developing the software, such as algorithms, and mechanical processes from the fly techniques.[9]
- The navigational skill of the foraging ant, which, using a range of memory, communication and physical tools, operates its own sat nav and is aware of its position relative to the nest, however far it travels – and in both featureless as well as cluttered environments. Useful parallels have been recently found between networks of ant colonies and human-engineered systems. One example is the 'anternet',

where a group of researchers at Stanford University found that the algorithm or rules that desert ants use to regulate foraging is quite similar to the Transmission Control Protocol (TCP) that is used to regulate data traffic on the internet. Researchers now believe these systems, developed over millions of years, will be invaluable in developing our own tools.[10]

It's worth noting that the nocturnal African dung beetle, *Scarabaeus satyrus*, is the only known invertebrate animal able to find its way using the Milky Way. The species was revered in ancient Egypt as a reincarnation of the sun god, and we can revere it now for its star-using skills.

Some species of cockroaches give birth to live young but only one so far, the Pacific beetle cockroach, is known to produce a protein-rich 'milk' full of carbohydrates, protein and other nutrients like mammalian milk. They feed this to their young as they gestate under their wings. Amazingly, some are suggesting this could be a new 'superfood' for humans, but it may be rather fiddly to 'milk' hundreds of cockroaches for a thimbleful. We may learn more from cockroaches, and from burying beetles which form family groups to help protect young by sharing gut protozoa that is essential for digesting cellulose from their woody diets. They lose this digestion tool after every moult as they grow. So, to ensure the young's survival, the adults stay with their offspring until they reach adulthood. There is an interesting analogy here with rearing humans, where we can enhance a child's resilience to infections through maintaining contact with mixed-age groups and ensuring exposure to microorganisms in order to build up childhood resistance.

Much can be learned from life in the 'invertosphere' – how bugs live, their habits, the way some species communicate and organise in vast communities and even between species. I am constantly amazed by bugs. How they manage their ecosystem, and how they find food and defend themselves. How they can adapt and survive and colonise the space they find themselves in. Understanding invertebrates better could unlock many of the answers to how we should and could live in harmony on this one planet we all call home.

## The lesson of diversity

Nature largely avoids uniformity and benefits from diversity and the mosaics of life and landscapes. It is the mix of species and interactions that makes an ecosystem work. Bugs are a critical part of that mix. A 2012 study found more insect species in just over one acre of rainforest in Panama than there are mammal species in the entire world.[11] They learn to fit their niche with incredible skill.[12]

Humans, however, have been spreading uniformity across the land via crops or grazing, creating one of humankind's major problems. This clearly separates us from the rest of nature. We create vast areas of land in homogeneous, monocultural food or fibre production. We nurture garden lawns that look the same the world over. Many would also argue much of our education system creates uniform thinking and robs children of the ability to be creative and think for themselves. We should learn from insects how to fit into and live with nature, rather than assume we are separate from it and can fix any threats through science and technology.[13]

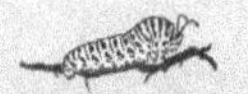

## Electric leaps

One striking tool that bugs use to travel has recently been uncovered in the small money spider. In 2018, Bristol University researchers were investigating the spiders' ability to reach huge heights for an animal less than 5 millimetres in length. They uncovered an astounding fact. To reach the heights they need, money spiders create an electrostatic force from the flower or surface they are aiming to leap from. This propels them far higher into the air than from purely muscle power and wind drag on the silk thread they shoot out. The process, called 'ballooning', uses atmospheric electricity and allows the spiders to elevate huge distances, and therefore the ability to traverse habitats and even continents. Many other invertebrates, including butterflies, are also known to use ballooning to travel.[14]

We also need to move around, and hard borders are not helpful. Migration allows for the mixing of genes and parentage, and maximizes the opportunities to ensure species' survival and adaptations. This is as true for us as the invertebrates, as it provides opportunities for helpful mutations. Invertebrates being more numerous and breeding rapidly can evolve new forms to adapt to their environment. The assumption that species, or indeed people, should be fixed to a specific place is not borne out by evidence. We should

### How much is a bee worth?

Policymakers like to think in terms of value, usually monetary. Yet viewing invertebrates in terms of financial gain is misleading as we could not really live without them. Yet it is worth noting that some have put a staggering monetary value to the 'services' bugs provide. One paper in 2006 put a conservative monetary value on what insect 'services' alone – not to mention all the other invertebrates – provide to the US economy. It was $57 billion (around £45 billion) a year.[15]

Another global study, using data from 90 studies and 1,394 crop fields in 2015, showed the value in pollinator services of wild bees at over $3,000 per hectare.[16] This work suggested that, in the UK, bees bring in more than £651 million a year to the British economy, which is apparently £150 million more than the royal family bring in through tourism.[17] There are indeed queens who are worth more than the Queen. But as so much of our food crops and flowering plants need animal pollination, it is clear that our wild invertebrate pollinators are critical in far more than monetary terms.[18]

see migration and mixing as a vital part of survival. A lesson we already know and yet we put up barriers and walls with increasing frequency and conflict.

— CHAPTER 2 —

# What bugs do for us

The truth is that bugs are exquisite in their evolutionary design and they should be able to exists for their own sake, not just for the value in service they give us. But they should be revered for all the vital roles they do play in keeping our only and shared home inhabitable. We just need to see them differently. Looking at a few choice examples of both iconic and totally obscure bugs, I hope to show you their key role in nature and why we depend so much on invertebrates.

## Beetles and worms turn waste into fertiliser and food

We would be knee-deep in manure and sewage (not to mention leaf litter) within weeks if bugs did not find organic waste an attractive food and habitat. With over five thousand species of dung beetles, we have an army of tiny waste contractors for every job. They are a keystone species involved in vital ecological systems including soil aeration, nutrient cycling and seed dispersal. Species exist on every continent except the Antarctic, and they are vital and often incredibly beautiful. Soils would be so much poorer without them.

Dung beetles seek out the fresh droppings of animals, then live on this animal dung and even lay their eggs in it. Some roll the dung into tidy balls, which they then use as food and a nest for laying eggs – their hatchlings have a ready-made breakfast. Others live in the dung or tunnel below it, drawing the dung down into the soil.

Many other bugs are vital removers, converters and consumers of waste, including flies and worms. We would find it hard to live around all the dead bodies were it not for the carrion beetles, rove beetles, flesh and blow flies, ants and wasps that feed, and lay their eggs, on dead flesh. Maggots may be revolting, but they are also fascinating and useful – without them, we would soon find the earth unliveable.

Worms deserve a significant mention as they, too, recycle waste materials in a way we could not do without and provide a vital function, which is biological, chemical and physical. They are found worldwide and break down both faecal and plant matter like leaves, freeing vital nutrients for plants and crops to use while also aerating soils. The complex interaction of worms with soil fungi and microorganisms is only just beginning to be understood. The worm is the farmer and gardener's best friend, but they are so badly affected by the use of machinery, artificial fertilisers, chemicals and the loss of organic matter in soils.

### The resilience of the moss piglet

Another ubiquitous water dweller is the tardigrade, which has the charming common names of 'water bear' and 'moss piglet'. Tardigrades are a unique, remarkable, tiny eight-legged bug. They are probably the most resilient animals known, able to survive extreme temperatures, pressures (both high and low), desiccation, oxygen deprivation, radiation and starvation, which no other life forms could withstand. They can basically shut all their systems down for years and reanimate when conditions allow. The tardigrade can survive in space and there might even be moss piglets on the moon, having been dropped there in a failed moon landing attempt in April 2019. They have survived all five mass extinction events and have been revived from a one-hundred-year-old moss sample from a museum. They also play a crucial role in water systems – as colonisers in new habitats and as vital food for larger species.

## Rotifers are expert water protectors

Many of our water and sewage treatment systems rely on bugs. Worms are an essential part of many water treatment systems, helping break down waste materials and aiding the filtration process, but there is a vast army of water purifiers we rarely see even in the wild. Take the tiny rotifer, whose

Latin-derived name means 'wheel bearer'. These are microscopic invertebrates that are critical in purifying water across the globe. They get their name from a circle of cilia – tiny hairs – on their head that look like a spinning wheel. They spend their days moving about in watery environments, eating the bits of organic matter, dead bacteria, algae and other smaller bugs that live or have fallen into the water. Their little crown of hairs direct the morsels of food into their mouth, which they digest, then excrete the unused parts.

By breaking matter up into even smaller pieces, rotifers help the microbes to further decompose this material. The nutrients that are released by this process then become available again for plants and animals to use. The work of the rotifers is not finished there. Through their activities in curbing filament bacterial growth, they may also help to increase the oxygen content in the water, making it more hospitable for other plants and animals to live and breed in. Like worms, rotifers are also a critically important food source for many different species, including larger animals like fish and birds.

Rotifers can be found in virtually every lake and stream, every pond, on bits of moss in moist forest, in the tropics and even in icy Arctic terrains. An almost guaranteed place to find them is in sphagnum moss at the edge of a pond or lake if you have a handy mobile microscope to investigate…

## The hummingbird hawkmoth's tongue

Invertebrates servicing the sexual needs of many of our plants is something most of us are probably aware of but know little about. Most of us have some inkling of pollination. There is the well-known example of the bee – the way they transfer the male pollen from one flower to the female parts of another to enable the plant to produce offspring, and in turn giving us crops like apples, peppers and almonds.

## Look out for the hummingbird hawkmoth

The hummingbird hawkmoth is an exquisite and large moth that flies during the day – happily for us as it is gorgeous to watch. It is a summer visitor to the UK and not always common, so it's a treat when you do spot one, not least because of its resemblance to the hummingbird itself. To encourage these beauties, many gardeners will grow the moth's favoured nectar plants, such as honeysuckle, jasmine or red valerian. Leaving the plant *Galium aparine* – commonly known by many names, including cleavers, goosegrass and stickybuds – which is too often considered a weed and so removed will help, too, as this is one of the moth's preferred breeding plants to lay eggs on. They also like to lay on lady's bedstraw, another type of *Galium* (bedstraw) plant, which produces bright yellow flowers and can easily be bought and grown if you would rather not have cleavers about. Always a good tip is to leave some of your garden messy and, of course, to avoid chemicals.

My first wild experience with the spectacular hummingbird hawkmoth in England was actually in my tiny twenty-five-metre squared London garden where I had let cleavers or 'bedstraw' run wild. With its wings beating at eighty-five beats per second (that is thirty-five beats more than a hummingbird) I could almost say my heart was beating at the same frenzied pace. I have lived in the same house for twenty years and not had such a visit. With no chemical applications and left to itself, a garden can become an amazing refuge for wildlife.

But how about the hummingbird hawkmoth? Their pipe-like 'tongue' is precisely adapted to reach the parts of a flower other insects cannot reach. By evolving this extremely long proboscis – strictly speaking, a sucking mouthpart – the moth can penetrate deep into tubular flowers to reach the rich nectar inside.

It is a masterpiece of engineering and design, and even more impressive when you realise its wings move so fast as they hover over flowers that they hum. To be able to move flexibly, it can use free-hovering flight, using a complex set of musclular, sensory and wing adaptations. After the hummingbird hawkmoth leaves the flower it has probed deep for nectar, it takes the dusting of pollen grains, carrying the male genes that have fallen onto its body, to another flower it is visiting. The pollen falls onto the female parts of the next flower and so fertilises the plant.

Such moth-powered genetic mixing, allowing sexual reproduction of the plant, is critical in keeping plant species fertile and vigorous. It is also a beautiful example of excellent design and mutualistic relationships.

Many other bugs (bees, flies, wasps, moths, beetles, ants and even mosquitos) that pollinate plants are usually visiting the flower to get at a food source – the nectar – and carry the pollen by default, though some species also use the pollen they collect too. Bees use it to feed their babies as it provides a vital protein source. A single bee can carry about half her own body weight in pollen.

These sorts of mutually beneficial relationships between bug and plant have evolved over millions of years. In many cases the relationship (and physical attributes that allow it, such as long tongues and complex flower structures) are incredibly elaborate, specific and unique.

## Wasps as plant defenders

Despite a bad reputation, wasps provide hugely important services – from pollination to pest control. Don't kill them because they are just as useful as bees and ladybirds and remove many flies you don't want. If you have a wasp nest and can't leave it, there are many ways to remove it without harsh chemicals, but it's wise to get expert advice.

I spent a happy summer in the late 1980s in Suffolk, England, breeding a type of miner fly as part of my pest management studies. The leaf miner fly is considered a pest as it lays its eggs inside leaves and, when the eggs hatch, the maggot larvae bore their way through the leaf, making visible white trails. In large numbers, this can lower fruit yields or make plants unattractive as the leaves are covered in white lines. It is a big problem for large glasshouse crops.

My research was to see whether we could rear a small wasp biocontrol in commercially useful numbers for use in glasshouse crops. The wasp is a 'parasitoid', which means it kills the fly by laying its eggs in the fly larvae, which then hatch into wasps and eat the larvae as they grow. I scoured

### Hunt for the leaf miner fly

Rebugging is not only about research. Leaf miner insects also provide a fun nature treasure hunt game to play for children and adults (I had fun looking for the hard pupae). Find a grass blade or leaf covered in the distinctive white squiggly lines left by the leaf miner larvae. Look for the wiggling bug still inside eating – you can often see their mouthparts scraping away at the plant matter. Or spot the hard pupa it has turned into before it emerges as a fly.

maize crops in the hot sun for signs of leaves with the telltale miner lines and tiny hard pupae metamorphosing inside into the adult cereal miner fly. I would then take these leaves and hatch out the flies in cages, provide the conditions and food plants to encourage them to lay eggs and introduce the wasps into the cages I'd built. It was a blissful time and I found enough fly pupae to start breeding before I left to go back to university. The short study showed you could breed these predator hosts in a glasshouse environment. You can buy such wasps to control leaf miner flies in your glasshouse and so avoid the need for insecticides.

## The ladybird pest controller

Many invertebrates can be a pain when it comes to food growing. Slugs, aphids, caterpillars, cutworms – all enemies

**Feeding ladybirds**

Many gardeners will work hard to ensure ladybirds and other useful predators are attracted to their garden. Gardening organically with no insecticides is vital as these will hurt the good bugs as well as bad. Planting flat-topped flowers such as yarrow, angelica, fennel and dill, along with common plants like calendula and marigold, will attract pest predators.

Oddly, it's also important to make sure you keep some pests, like aphids, as a food source to attract ladybirds and other beneficial wildlife. And keeping moist dark spaces where the adults can overwinter, like old hollow stems and logs will help.

of farmers and gardeners. Billions are spent each year in chemicals and research to kill them off or keep them out.

But bugs can be a huge friend to the gardener and farmer when given the right conditions. One common example is the ladybird, which is a voracious carnivore and can keep aphid numbers at acceptable levels. The fierce-looking larvae of some species – black armour with red stripes – suggest their predatory nature and they can eat up to fifty aphids a day.

Gardeners may also buy tiny nematode worms (bought in the form of eggs, which hatch when added to water) to control slugs and snails. Such 'biocontrols' (see the section 'Bugs as pest-management strategies' on page 42) can be self-sustaining – because the bugs breed future generations of the control – harmless to the public and cost effective. The economic value of naturally occurring biocontrols is hard to assess – you would have to remove them to calculate their contribution – but is likely to be billions of pounds annually.

## Bugs as pest-management strategies

Bug-based biocontrols are increasingly viewed as useful tools in farming and could become big business, playing a vital role in integrated pest management strategies. This is not new, but against the competition of cheap, abundant and effective chemical pesticides, it has been hard for researchers to devote money to testing and trialling these kinds of tools. But maybe their time has come as society demands less use of harmful pesticides, and as chemicals have started to fail because pests and diseases have become resistant to them. The use of invertebrates in pest and weed management can and should be greatly enhanced.

Globally, the use of biological controls can avert disaster. This was the case with the cassava mealy bug, which by the 1980s was decimating crops in Thailand and causing rapid deforestation in neighbouring countries who increased cassava cropping to meet demand. When a parasitoid wasp (which lays its eggs in the mealy bug young, which then hatch into larvae and eat their host) was released, 'outbreaks were reduced, cropping area contracted and deforestation slowed by 31–95% in individual countries.'[1]

Inevitably, given our instinct to be in control, coupled with the complexities of nature, we have made mistakes with biological controls. The deliberate introduction of the cane toad in Australia in the 1930s to replace the toxic arsenic, copper and pitch pesticides in the control of cane beetle pests was a costly mistake. The toad very rapidly became a hugely harmful invasive toxic species, damaging wildlife across large areas and having little impact on the cane beetle. Recently, a biocontrol toad solution may have arrived in the form of water rats. Scientists have observed the rats killing and eating the toads, apparently carving out their organs with 'surgical precision'. Nature finds a way.

Yet, when carefully applied with expert risk assessments and as part of an integrated strategy, bug biological control can deliver major environmental and economic benefits. Since the late 1800s, it is estimated that over 200 invasive insect pests and over 50 weeds across the globe have been completely or partially suppressed through biological control.

## Maggot therapy and medicinal leeches

The healing power of bugs is also little known to most, but they have been used in medicine for thousands of years. It is recorded that the maggots of flies were successfully used in the treatment of wounds during the American Civil War, and records of their use go back to ancient times in human and animal wound treatment. Though much is historical, it is worth noting that some bugs are creeping back into modern medicinal use.

It's their ability to eat rotting flesh that is key. Live, disinfected maggots (often larvae of the common green bottle fly) are put onto non-healing skin and soft tissue wounds of a human or animal to clean out the dead tissue. They can disinfect the area, too, as they have antimicrobial properties. This 'maggot therapy' is cleared for use in medicine today for certain wounds in the US and, apparently, patients not only accept the treatment but do well with it.

Honey has been used in medicine for thousands of years, which given its antimicrobial and anti-inflammatory action, is not surprising. The ancient Egyptians used it for dressing wounds and the evidence suggests that it can help burns heal faster.[2]

Leeches, a form of segmented blood-sucking worm, are also making a comeback in medical care after being abandoned for a century. They were used in ancient times and more recently in the Victorian era to draw blood from patients. Women would paddle in rivers, allowing leeches to latch onto their feet before plucking them off into boxes to sell to doctors or pharmacists, who kept them in ornate jars.

Medicinal leeches stop blood clotting, which allows them to keep sucking and, as they use an anaesthetic to numb their host, it is painless. These days leeches are used to treat diseases in joints, veins and in microsurgery, and the anti-clotting agent is now manufactured and used to treat some blood illnesses. If you encounter them in nature, there is little risk. My young son once had a leech attach itself to his foot after a cold lake swim in Canada. His younger brother immediately went in to find one that would suck his blood too. We were so happy and fascinated, and I managed to take a photo before prising the leech off my beloved boy.

## Bugs as a delicious food source

Invertebrates are a core food source for many animals. Without them many bird species would not exist. But many humans, too, rely on bugs as a direct source of food. The Food and Agriculture Organization of the United Nations has estimated that around two billion people eat bugs directly – a practice called entomophagy. They catch or rear insects on a regular basis benefitting from this cheap, high-protein food in their diet.

Companies in more affluent countries, where eating bugs has long since been dropped, are working hard to develop insect-based products that consumers may find acceptable, like bug crisps or cricket flour. Insects are already reared for fish feeding, but systems are being tested on a much larger scale to produce cheap insect protein in industrial quantities for feeding livestock like chickens and pigs.

I once bought some cricket flour online to try it out. The smell reminded me very much of my days as a student, chasing cockroaches back into their containers in the lab. Not very appetising to be honest. I produced very sugary, gingery cookies to hide the smell. Most of my fellow office workers were happy to try them and we all found them pretty tasty. A successful small trial.

A friend of mine was inspired by the copious tasty cricket-based street food in Thailand to start breeding her own back in England. She now has over seventy thousand crickets chirping in her study. Though she cannot yet legally sell cricket flour from her Six Legs Farm, she has managed to sell the manure they create, called frass, as garden fertiliser.

In the US, the sale of edible insect food products is legal but, in the UK, currently the only legal form of insect rearing is as feed for aquaculture. Imported products made from insects in supermarkets are legal, but the rules are now changing to allow for commercial UK developments in insect flours, whole insect snack products and in livestock feeds to replace protein crops like soybean meal.

In early 2021, the European Union approved mealworms for human consumption and is expected by the insect industry to start endorsing other insects, like locusts, crickets and grasshoppers, as being safe for human consumption. There could be huge new opportunities for the mass production of a range of insect dishes to be sold across Europe for the first time as well as insects in livestock feed. Many millions of pounds are now being invested in maggot factories.

While I think major ethical, safety and environmental questions still need to be answered in relation to large-scale commercial production, there is no doubt insect protein could replace a proportion of the highly cruel and unsustainable livestock systems we currently use. But insects should be considered sentient animals, capable of experiencing different feelings such as suffering pain, and so they should be reared and slaughtered using humane methods.

Additionally, what these billions of animals, however small, are fed on really matters – waste food material or even animal waste may be diverted from other important uses, such as a soil nutrient or improver. This could create problems and additional demand for artificial fertilisers. As we count the cost – to health, greenhouse gas emissions, animal cruelty, biodiversity to name but a few – of our current livestock rearing practices, insect eating is being viewed as a cheap, ethically easier option. But let's not jump into another large-scale disaster with insect rearing.

## Do bugs feel?

Humans have a sense of superiority over the rest of the natural world because they assume it has little sentience – the cow or the donkey is less 'feeling' than us so we can use and abuse it. But campaigning by organisations such as Compassion in World Farming (CIWF), among others, has led to the establishment of legal definitions on animal sentience that will continue to challenge these unhelpful assumptions.

Do invertebrates have feelings? Does this affect their ability to fulfil their needs? And should this affect how we treat them or learn from them?

Scientists use the term sentience when trying to explain the concept of feelings. Sentience is the capacity to feel, perceive or experience subjectively. A sentient being can show awareness or responsiveness. Having senses makes a living body 'sentient': able to smell, communicate, touch, see or hear. Clearly invertebrates can do all these and more.

There is a definition that suggests non-sentient living organisms are those without a centralised nervous system. This would include bacteria, archaea, protists, fungi, plants and certain animals. According to the science, they lack nerve structure or physical structure complex enough to allow for the possession of consciousness.

The long campaign for recognising animal sentience by CIWF and others had a breakthrough in 2009 when the European Union Lisbon Treaty stated that 'the Union and the Member States shall, since animals are sentient beings, pay full regard to the welfare requirements of animals.' Giving a legal definition of animal sentience is leading (albeit slowly) to better regulations for care, transport and slaughter of farm animals.

Cephalopods (including octopus, squid and cuttlefish) are invertebrates, but they are often considered to be more 'advanced' than others and are now recognised in law as sentient. The science now suggests they feel pain – a complex area for investigation. In 2013, animal welfare legislation in the European Union gave protection for the first time for cephalopods used in scientific experiments. This means there was sufficient scientific evidence that they experience pain and suffering. This also means there is ongoing work to ensure they are protected in all countries and, for all instances, including their care, housing, feeding and slaughter for consumption.

But there is still a huge question mark over most invertebrates which are not protected. Millions are already reared for fish feed, but that will be dwarfed by growth for livestock feed, particularly poultry and pigs. The US insect feed industry alone has been estimated to be worth nearly $2.4 billion by the end of 2029.[3] This will have huge implications for trillions of animals that will be reared and slaughtered every day.

Having absent or inadequate welfare regulations for what could be trillions of invertebrates used daily for research, feed and food is unacceptable. Legal definitions and laws covering their care and slaughter are now well overdue. We can see from current research that they exhibit many aspects of sentience including communication, touch and feel, and there is no evidence to suggest that they do not feel pain of some sort. This should mean, at the very least, the precautionary principle applies, leading to welfare rules for all invertebrate rearing and slaughter. Whether we should be rearing them in vast quantities as with industrial farming of mammals and poultry, is another question.

---

Considering the many critical ways invertebrates matter to us and the data suggesting they are in serious trouble, we clearly need to act. But we should act not only for our own needs. Invertebrates have a right to live in an unpolluted world, free from dangers, free to live their lives as naturally as possible. But even if you do not subscribe to that view, there are other, excellent reasons to care about bugs. But the most fundamental reason is that without them, life on this planet would not really be possible. We need to learn what we're doing wrong and how to put things on the right track.

— CHAPTER 3 —

# Rewilding by rebugging

What is rewilding? Basically, it's the attempt to recreate the natural ecological systems that once covered our landscapes – woods, rivers, wetlands – and trusting nature to look after itself, perhaps with some help at the start to fix the most broken pieces.

Many rewilding projects are large in scale, to allow nature to really do its stuff without interference and pollution from us. It is about vast estates and landscapes, large herbivores or carnivores and huge decisions made by distant landowners or institutions. These are invaluable. But is not always about completely removing people – after all, humans are part of the natural world. Instead, we need to find new ways to live while reconnecting with the ecosystems we live in, creating a richer world in which people and nature can thrive together. We can live alongside more bees, worms and flies, and I believe there is a benefit to taking the debate on rewilding down to the tiny scale of some of the smallest creatures on the planet.

Invertebrates are core to any rewilding project: ideal foot soldiers for the cause at every level as they travel, adapt and multiply so brilliantly. And, aside from farmed honeybees, silk moths and biological control agents, almost all the invertebrates we encounter, wherever we encounter them, are *wild*. They may be there because we created the environment for them, but they are not domesticated or tame – or even that interested in us.

## How does rewilding help bugs?

Rebugging is looking at the ways, small and large, to nurture complex communities of these tiny, vital players in almost all the natural and not-so-natural places on earth. It means conserving them where they are managing to hang on, and restoring them where they are needed as part of a rewilding movement. And it means putting bugs back into our everyday lives, our homes and where we play and work.

But what does 'good' look like for the bugs? We need to better know what the 'perfect' habitats and conditions would be for bugs to thrive: the baselines against which recent losses occurred. We can't tell what the true losses are as we don't know what was there before people arrived, or even a hundred years ago. But how exciting to discover more new insect habitats through rebugging, as we let nature make its way. Even rewilding a relatively small area can create something akin to the original habitats of the invertebrates, and we will discover so many intriguing aspects in the process. Rewilding projects are already throwing up some challenges to our previous knowledge about their favoured habitats as species take to a habitat in a rewilded area that we had no idea they liked.

Invertebrates played a fascinating role in the inspiring story told in *Wilding*; a book by Isabella Tree who, with her husband Charlie Burrell, decided to rewild their 3,500-acre West Sussex estate in the UK during the 1990s after years of trying to make conventional farming work.

I was thrilled to read about the many invertebrate species that arrived back on their estate, sometimes in astonishing numbers including ones that had almost disappeared from the UK such as the purple emperor butterfly.[1] The real intrigue was how they found species known for living in one type of habitat in the UK, using a different one entirely at Knepp, suggesting that some insects are living in some habitats only because they cannot find their natural one.

The purple emperor, for example, was previously considered a woodland species, but it turns out they love a scrubby habitat. This starts to get my pulse beating fast as you consider all the assumptions made in conservation about 'ideal' habitats for bugs – in research papers and conservation texts – about what they like to eat when, actually, they could have broader tastes. And therefore, what we could achieve by rebugging and bringing in more diverse habitats everywhere, including in urban places.

Learning from the rebugging experience can be hugely rewarding. Anyone who has read of the trials and joy of Isabella Tree and Charlie Burrell in rewilding their large estate will have seen the vital and complex role of invertebrates. The learning curve rewilders undergo is a fundamental and often extraordinary part of the process.

For instance, Tree describes a difficult period when they were in the middle of rewilding their estate at Knepp Castle. A particularly strong pioneer weed (pioneer species are plants that are the first to grow on barren land) called creeping thistle took over with 'breathtaking speed' in 2007. It has other rather apt names including 'lettuce from hell thistle' and by 2009, the weed had taken over huge areas of the estate. Tree and Burrell were keen to let nature take its course – the point of deep rewilding – but the invasion was a huge worry for their neighbouring farmers and threatened their grant funding.[2]

Yet nature did have a solution in the form of the painted lady butterfly, which came over that year in staggering numbers.

Possibly the greatest butterfly migration on earth, up to 9,000 kilometres round trip, was flying over the Knepp estate, laying eggs in their favoured host plant – the thistle. Spiky black caterpillars ate through the fields of thistles leaving little left to worry about. The boom and bust of nature – where the ability of insects to move and reproduce fast can be crucial – was demonstrated in stark clarity in this rewilding project.

## Bringing back lost species

Which animals belong where is a fascinating issue in rewilding. It can involve reintroducing a species to re-establish it or to boost numbers of a native animal or plant at risk of going extinct. Or it can be about recreating an ecosystem that has got out of balance, such as a flood plain that needs the plants and animals back to slow water flow.

Would we want to bring invertebrate species back into countries and regions that have lost them? The removal of keystone species – a species that is fundamental to the existence of a particular ecosystem – can be catastrophic for a wild ecosystem, but reintroduction can work in unforeseen ways. The reintroduction of wolves to Yellowstone National Park in the western US, created unexpected and positive results for the park ecology. When wolves were removed from the park seventy years ago, elk overgrazing became a problem and only resolved when the wolves were reintroduced, and so elks were naturally managed better. But there was a further impact: beaver populations grew now that their willow trees were not overgrazed by the elk. This created new fish and water invertebrate habitats, which then influenced other species feeding on the bugs and fish.[5] Everything is connected, and while many focus on the furry vertebrate species, we need to recognise and nurture the bugs, too, as vital parts of the arrangement.

## The wood ant in the Caledonian Forest ecosystem

The wood ant shows us what can be done when bugs, however small, are in charge. Two species of wood ant that are on the International Union for Conservation of Nature's (IUCN) Red List of Threatened Species are critically important native inhabitants of the Caledonian Forest in Scotland. So boosting their numbers is a priority. As social insects, wood ants live in colonies which can number up to half a million individuals, mostly worker female ants. The queen can live up to fifteen years and the nest they build is intricately designed to manage heat, air flow and water. I recall coming across a large wood ant mound when walking with my children in a Swiss forest. We witnessed an extraordinary level of activity, gathering nest material and food on the mound, and we could hear the ants working hard.

These ants are extraordinarily versatile. They hunt for prey both on the ground and in trees, but they also farm aphids, getting the aphids to release droplets of honeydew, a food that is rich in sugars, acids, salts and vitamins, which the ants eat. In return, the ants protect the aphids from predators like beetles and parasites. They also have a symbiotic relationship with a species of earthworm, *Dendrodrilus rubidus*, as conditions within the nest are very worm friendly, with an abundant food supply. In return, the worms stop mould and fungal growth. The nests provide a habitat for other species, such as the beautiful rose chafer beetle, whose larvae feed on plant debris inside the nest, and allow vital nutrient recycling once the nest is abandoned.[3]

The ants also provide food for many forest species, including the rare capercaillie (a large woodland grouse) and they play a key role in seed dispersal. When wood ants are removed from a woods, the ecology of the woodland system can become unbalanced.[4] Other species populations – kept in control by the ant predation or activities – get out of control and can cause major damage, for example, the pine looper moth caterpillar in North American forests, which can strip a tree of needles. The importance of these tiny ants – a keystone species – is only now being properly recognised and the Caledonian Forest project is providing a crucial tool for this research.

It's worth giving ants credit for forming the largest known cooperative societies – 'super-colonies' often across hundreds of miles. One super-colony of Argentinian ants was found in Southern Europe that is estimated to stretch 3,700 miles across and have 33 ant populations with millions of nests and billions of workers. Furthermore, scientists have found that even across continents these super colonies of Argentinian ants may be related which would explain why they are not aggressive when mixed. Amazing creatures.

Beavers are also being reintroduced into UK river systems, leading to new habitats, more diversity, and even floodwater management and boosting green tourism. Sometimes iconic species can be hugely important for building public support for conservation, but also can help fund projects through carefully managed tourism.

But what about invertebrates? Rebugging could allow species lost to an area to be introduced successfully and this is indeed happening.

Given their size and ability to produce numerous offspring quickly, invertebrates have the wonderful ability to recolonise far more quickly when they spot the opportunity than larger species. Just take the aphid, which can produce five to ten offspring every day. The African driver queen ant can produce an estimated three to four million eggs a month. And they do not need so much careful handling as, say, a wolf. However, it makes sense also to focus on protecting the native bug species that are still in their habitats, but are just hanging on in pockets of scrub, hedgerows or small woodlands, and even urban parks, where once their habitats would have been far more widespread. And they can help rewild the small spaces as well as the big ones.

## The school of rebugging

We need to learn to move at speeds that work for nature and learning to wait is another lesson. 'A marathon with a sprint start' is how Alastair Driver, director of the charity Rewilding Britain, has described rewilding. You may often need to do something significant at the beginning, such as the reintroduction or removal of a species or habitat to kick-start the rewilding, then leave nature to take its course.

### *Farmland*

Farmers introducing 'wild' spaces to their land – such as native wildflower strips, ponds or small woodlands – see multiple benefits as biological controls of pests and even weeds enter the system. Many natural enemies of pests rely on flowering plants for a part of their life cycle and, when introduced, they can reduce crop damage to economically acceptable levels. I have met many organic, nature- and conservation-focused farmers who are well versed in these techniques, and there is a rapidly growing network of nature-friendly farmers learning and researching what can work better for the wildlife in their fields.

Agroecology is a form of farming where nature is a central tool for production. Agroecological farmers, like organic farmers, nurture beneficial invertebrates to manage nutrients, pests and pollination. Many studies now confirm the benefits of careful planting of flowers species in or around a crop. One project has looked at the best way to attract natural enemies of two harmful apple tree aphids on the edge of Mediterranean apple orchards. With the careful selection of flowers for the edge of the orchard they could attract five different predators of the aphids. They lured parasitoid wasps in, too, which as we know, lay their eggs in aphids who subsequently die when the baby wasps hatch.[6]

Food plant pollination is also boosted by rewilding. One study demonstrated that over a four-year cycle you can get more flowers fruiting and a greater weight of berries following the planting of wildflowers in field margins, which attract pollinating wild bees and hoverflies.[7] This means more pollination, possibly more pest control, and so more fruits and seeds being produced. The profits from these increased yields would outweigh the costs of planting wildflower strips, including any land lost and the preparation and maintenance.

Some would argue this is not 'rewilding' as such, and it has its limitations, especially in large fields, and may not increase overall numbers of beneficial insects on farm. Yet for me, this is rebugging with co-benefits for growers.

*National parks*

Unique habitats in national parks can be vital for some of our rarest invertebrates like the swallowtail butterfly or the glow worm, which is actually a beetle not a worm (the wingless females climb up plants and make their abdomens glow to attract males) and is nurtured in the South Downs National Park in southern England.

The Caledonian Forest in Scotland is a unique patch of ancient forest managed by the Trees For Life conservation body.[8] Truly a fantastic refuge for bugs, they are being developed as a major rewilding project by restoring to a wilderness an area of 2,500 square kilometres. To kick-start the work, they have planted over a million trees and far more will come naturally to reforest the area. Invertebrates that have already returned in this rewilded refuge include the strawberry spider, wood ant, black darter dragonfly and the pearl-bordered fritillary butterfly. The project is about the whole system: regenerated in ways that nurture complex ecological habitats. It is also open to the public so visitors can be excited by the nature on view and even inspired to get some rebugging in their own gardens or community – through tree or shrub planting, for example. Maybe local communities will be inspired to join together and petition local councils to create a 'tiny forest' or orchard (see chapter 4) with thriving wood ant nests in every town.

There is a hot debate now as to whether these publicly 'run' parks should be entirely rewilded and their management left to nature. Many people question the idea that these parks represent the best we can do and that the landscapes created are somehow culturally significant or even the right mix of nature and biodiversity.

It has been suggested, for instance, that England's largest national park, the Lake District, should be allowed to rewild with fewer sheep. The sheer number of sheep grazing, often fuelled by subsidies, inhibits the growth of natural vegetation, such as shrubs and trees, as they would if left to nature. It is now designated as a UNESCO World Heritage Site, but that means keeping the unwooded landscape largely as it is now with the sheep and their associated farm systems, culture and communities. It is not easy to see where this debate will end, but in the meantime many parks do provide a valuable refuge for invertebrates and some experiments in rewilding are already underway.

Critical to keeping places wild and protected will be helping people to have a stronger relationship with nature. Making public access safe and easy in rewilded space will help create a movement for rebugging. Great wilderness parks such as the sixty-three federally designated US national parks present a whole other level of invertebrate opportunity. As these areas are managed by government bodies largely for wildlife, rather than farming or other purposes, they can be described as wild – and over 80 per cent of the areas

**The swallowtail butterfly**

When my boys were young, we spent a family holiday on a great UK national park – the Norfolk and Suffolk Broads – on a boat slowly motoring through a fantastically preserved watery heaven. There, we saw the rare swallowtail butterfly, one of the most spectacular of the UK's butterflies, with a wingspan of up to 9 centimetres. It is currently only seen in the Broads as this is the only place that milk parsley for egg laying and caterpillar feeding remains. We managed to observe two phases of its life cycle – the larvae on milk parsley and the adults feeding – an invaluable lesson on metamorphosis for my two young boys. This national park is one of Britain's largest protected wetlands and third largest inland waterway, and it holds many more of the UK's rarest species, such as the white-clawed crayfish and the Norfolk hawker dragonfly.

involved are managed as wilderness. They maintain some of the best habitats, perfect for invertebrates to thrive. This is an extraordinary asset, but one which compares dramatically with other land management in the US: the empty prairies and often car-filled cities, where insects and other invertebrates are subject to massive pressures from industrial farming, pollution and development.

Take the sub-arctic Denali National Park and Preserve in Alaska where there is an abundance of invertebrates such as bees and flower flies. People visit this park to see the grizzly bears but the other fur-covered animals should also gain attention. Alongside the flies, the bumblebees are critical for pollination and they have recently found a new species of bumblebee in this park – always an exciting moment. These are keystone species and the Denali park's grizzly bears, caribou and wolves would not survive without the bugs because they all need the wildflowers and shrubs for their food or the food of their prey. The grizzlies in particular need the bees to pollinate the blueberries, one of the bear's main foods. As we know, honeybees are under threat globally, so it is vital that we protect the other pollinators like bumblebees so they can pollinate both wild plants and farmed crops.

Wildlife parks do have threats such as the pressure of visitors, especially at peak holiday periods. Other dangers respect no boundaries, for instance, climate change, illegal hunting and invasive species. But these places provide a fantastic way to conserve bugs in their natural world and to show what they can do.

*Rewilding in your garden*

What can you do in response to gain a stronger connection to nature? Small-scale rewilding can happen in your back garden and in your local park, as well as on a large estate or national park. It can also be about what you buy and eat. I want the large-scale rewilding, too, where minimal interference by humans lets nature seriously take its course. And I want everything else in between.

## A good start

The growing public appetite for all things invertebrate and their role in our lives has been wonderful to observe. From

## Bugs in your garden

I never really considered the patchiness of my rough lawn until I spent a lazy summer's day in a hammock and saw how many tawny mining bees were seeking bare ground to make their nest holes in. I knew that bee flies – one of those parasitoid species – lay their eggs in mining bee eggs, so I was not surprised to see a few of those later in the week. The golden, fluffy fly hovers in front of an open nest cavity and flicks its eggs inside with great precision and skill – the circle of life playing out in my tiny garden in polluted North London.

We all can relate to the joy of seeing bees and other pollinators return to the garden when flowers are reintroduced, or beetles settling into a log pile left alone to rot. Learning how the simple act of leaving a patch of lawn uncut can deliver a new system complexity and opportunity for investigations and observation.

The Organic Growers Association recommends mixing flowers with vegetables in your garden or vegetable plot to attract beneficial wildlife, such as caterpillar-eating birds and aphid-eating insects. They advocate applying prevention and cultivation methods, rather than chemical controls, to manage pests and diseases.

being fearful of insects, a growing proportion of the public has turned into bug lovers. My friends, family and colleagues have made bee hotels, planted wildflower patches and left their lawns uncut. They send me photos to identify a bug such as a swollen-thighed beetle and even an exotic shield bug found in imported green beans. It has stretched my ID skills somewhat.

It is true that there have been volunteer communities of dedicated bug observers for hundreds of years, documenting and drawing bugs so we have their invaluable records and illustrations showing species changes over many years. But in the last decade or so there has been a huge groundswell of interest stimulated by various initiatives, such as citizen science projects like the UK's Bee Cause campaign and the Western Monarch Count in the US. Programmes, too, like BBC *Springwatch* have been invaluable in generating public excitement in the UK about the variety of invertebrates, their life cycles and roles in nature. I see a level of interest that was absent ten years ago, which is creating more bug lovers, more buzz and more people wanting to rebug.

Innovative campaigns by Buglife and other organisations have helped raise awareness and saved vital wildlife habitats and species. The new and often youth-led Climate Strike and Extinction Rebellion events across the globe have also been transformational, featuring the loss of insects and nature frequently alongside the climate crisis.

### A superhighway for invertebrates

You may not realise it, but you may be on a bug motorway that needs some repair. UK organisations have been creating a fantastic map of insect 'superhighways', called B-Lines, to be filled with wildflowers and protected so that insects and other wildlife can move freely across the UK to find new habitats to recolonise. The B-Lines need strong support, from everyone including communities and landowners to local authorities, so they can create new or restore native flower havens for bugs. You can add your pollinator projects, whether small or large, to the Buglife maps (https://www.buglife.org.uk/our-work/b-lines).

But the joy of rebugging is that you can do it almost anywhere. Give people the chance to act and to encourage some bees, or even hummingbird hawkmoths, in a green patch of land, and you can start to change hearts and minds. From a Costa Rican municipality giving bees citizenship to an amazing three thousand food-growing spaces making space for nature in London, it is possible – and it is happening. Urban greening is not only possible but crucial. The 'rebugging' title of this book was inspired by another, recent book *Rebirding: Rewilding Britain and its Birds* by Benedict Macdonald, who argues that to have more birds around, larger mammals must

## Rebugging actions

In your family, street, village, community, work, school, university, church.

If you do one thing:

- Talk about bugs to everyone in your community, with friends and family.
- Show your love of bugs by sharing pictures of invertebrates. Take a picture of invertebrates you spot and share it on social media – the photos do not have to be perfect (see the box, 'Smartphone bug joy').
- When flying ants or wasps arrive, spread the love for them and counter the hate.
- Tell children you know why bugs matter. Find them a bug whenever you can and get them to talk about what they see and think. Use National Insect Week, Buglife and other great resources for inspiration and activities.

If you can do a bit more, try one – or more! – of these ideas:

- Support local charities to grow bug-friendly plants and parks. Join your local park friends' group and ask them to replace pesticide use with alternatives (see Pesticide Action Network: www.pan-uk.org) and to grow flowers and let lawns grow.

- Grow flowers on bare soil in your area, such as around trees or on road verges.
- If your workplace has outside spaces or an accessible roof, fill them with flowers to encourage insects. Ask your colleagues to join in planting activities and explain that 'weeds' are food for bugs and birds.
- If you are at school, college or university, call for flowers, fruit and vegetables to be grown on and in buildings, and on the grounds. Ensure there are ongoing resources to manage the plots – you do not want it to disappear after you leave. Setting up a bug or insect club would help ensure your legacy lives on.
- Join in surveys like the Great British Bee Count and if you see anything that looks rare, check and send details to groups who log sightings.
- Talk to your local garden centres and retailers – ask them what they are doing to promote chemical- and peat-free gardening.
- Write to your council to ask them to plant more flowering plants, cut pesticide use and not mow or spray weedkiller on road verges, roundabouts and other urban spaces when in flower.
- Email your MP to tell them to act for bugs and whether they fail to respond or respond well, go to their monthly surgery to ask for action.

- Check your pensions and investments are not supporting pesticide manufacturers or companies that are known to be deforesting. There are ethical investment companies who can help.

be allowed to do their work and re-engineer the landscapes. Letting nature heal itself and letting it get messy is key to a revival in birds and other species. If we can use the lens of birds and beavers to understand rewilding, we should also use bugs.

## An important word on the downside of rewilding and rebugging

My hope in writing this book and offering these ideas for rebugging is to make it incredibly appealing, but rewilding campaigns have also raised fears in farmers and land managers that their existing ways of life will be swept away. What about the land we do have to work – for food-growing and other vital activities, and the cultures associated with it?

Farmers, a community I engage with in my job, have inevitably found rewilding debates worrying and difficult. They want to produce food or other crops and run successful businesses while being rewarded to look after the farmland as best they can. Being park keepers or rewilders is not what many signed up for, though I know many do much to protect nature and the landscapes they are part of.

Debates on reducing meat consumption, planting more forests and imposing conditions on farming to allow nature to flourish have been heated. It is even harder for the smaller

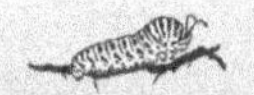

## Smartphone bug joy

The advent of the smartphone camera has been sheer bug joy to me. Admittedly, many of my photographs are taken on my handkerchief-size garden in London, but I now have a large and random collection of photographs, from dramatic crab spiders eating bees to cuddly bee flies, and from glorious southern hawker dragonflies to large wood ant nests. When I saw for the first and probably last time, a hummingbird hawkmoth in my scrubby patch, I got slightly over-excited… I still get huge pleasure from looking at these photos whenever I can.

This personal reflection is given as encouragement for anyone to do the same. Cameras on smartphones work magic, even if you are not an experienced photographer, and they are very handy if you are up for being a citizen scientist and helping to survey bug populations and find vulnerable bugs.

farm sectors, squeezed so badly by the food industry paying ever lower farm-gate prices (the price the farmer is paid as opposed to what you pay in the shops).

We need a revolution in food systems and to drastically change where the value goes in the supply chain so that farmers are properly rewarded. But there are helpful changes happening already. Farmers are being asked, and paid, to grow trees, to protect nature and provide other public benefits like locking carbon in healthier soils and protecting specific landscapes. A new Agriculture Act finalised in 2020 in the UK is designed to reshape farm support so farmers are paid for these kinds of public 'goods'.[9] This is my policy world and there are no easy answers. The pressures on land are going to explode as the climate and nature emergencies require increasing levels of action and coherent policies.

*The other side of bug life*

This may not fit perfectly with the core narrative that we do need to rebug, but we can't do it everywhere and with any bug. It would also be misleading to neglect the harm some bugs can do to humans and the immense damage invertebrate interactions can inflict on individuals, societies and on our infrastructure. Bugs have no intention to do harm, obviously, unless they are protecting themselves, their offspring or colony, but when nature is out of balance, invertebrates can do huge damage to nature.

As vectors of diseases like malaria, sleeping sickness and salmonella, bugs can cause many millions to fall ill, as well as hundreds of thousands of deaths every year. They can even stop whole armies – malaria spread by mosquitos was considered one reason why the twelfth-century Mongolian conqueror Genghis Khan did not manage to invade the whole of Europe in his attempt to expand his huge empire.[10] More recently, I recall the nasty but vital tablets I had to take while working in Ecuador, which, back in the 1980s, was a country with a high malaria risk.

More moderate but still a public health menace can be the stings and bites of invertebrates the world over and for those with allergies, they can be potentially fatal.

My academic studies touched on one of the biggest ways through which invertebrates cause damage by harming crops and stored food, from the flea beetle reducing oilseed yields to mass locust invasions wiping out a whole region's crops. When a crop fails, it can cause immense economic harm and food insecurity in low-income regions.

Yet this is a complex interaction and we are often to blame through the way we have developed super-vulnerable farming systems. Less than 0.5 per cent of all known insect species are significant crop pests, but a fifth of all crops are estimated to be lost to them annually both in the field and after harvest.

We spend billions each year drenching fields, forests and orchards with chemicals to stop bugs and nematodes from harming crops. We create the perfect place for bug pests to thrive. Crops selected for their size, high yield and nutrients all put together like clones in vast fields with few breaks or habitats for predators of the pests. It is a plant-eating bug's dream. Livestock bred for high yields in huge, closely confined herds are also uniformly vulnerable to parasitic bugs like liver fluke – a form of worm that takes root inside internal organs – or blow fly maggots, which infect sheep fleeces and can cause major loss of appetite and condition.

To tackle these issues, many farmers are increasingly turning to agroecological approaches, pioneered by organic farmers for decades, including more rotations, the use of bugs as biocontrols and rearing more hardy, native breeds in lower numbers. Such approaches need to be the norm rather than exception.

Damaging though bugs can be to us, their value far outweighs the harm. Yet we are making life extremely hard for invertebrates and are losing species, diversity and abundance on an alarming scale. We need to bring them back: rebugging on a small to large scale. For that we need to promote their value and help future generations to retain their youthful bug love and value the role of bugs.

— CHAPTER 4 —

# Parks and recreation: rebugging your world

Local green patches can really surprise you. I am fortunate to have a wonderful community orchard close by and I visit often to see what I can spot. For the first time in my life, in 2020, I came across a red-belted clearwing moth. I was following a speckled wood butterfly and as it settled on an apple the clearwing moth caught my eye with its distinctive black-bordered but clear wings and a thick, dark body with a single red band. I could not stop telling everyone I knew and sharing the photos I managed to take.

During the 2020 Covid-19 England lockdown, I found huge pleasure in taking far more time during the day to explore the local green spaces and also search for what is finding a home or food in our tiny patch of garden. Alongside hundreds of snails, I have counted over ten bee species and many smaller leafhoppers and shield bugs, weevils, wolf and jumping spiders, caterpillars and hoverflies, and even a large yellow-striped ichneumon wasp. My garden hums from March onwards.

Rather than seeing urbanisation as something that accelerates the loss of nature and wildlife, it should be a vital tool for both helping conserve biodiversity but also, crucially, connecting people with nature. And research shows that it is the amount of green vegetation as well as the variety of plants across different areas that are important in terms of the number and variety of invertebrates you find.

## Urban rebugging

Most of us will be able to find a patch of greenery nearby. It could be just a windowsill with pots, a garden or yard, a small grassy verge, a pocket park or, for the incredibly blessed, a huge park or the countryside. For those living in urban settings with no garden, parks and green spaces are how they experience the outdoors. They are also vital for the bugs. These spaces will be full of bugs or an ideal site for a bit of rebugging.

These green spaces are precious. They are vital not only for human recreation and well-being but for protecting wildlife and, when it comes to the smallest urban spaces, they are often best suited to the invertebrates. We need all these spaces and more to reverse the big decline in invertebrates as they can provide eating, nesting and breeding sites, and corridors for travel.

Recent investigations have looked more closely at the role of urban and recreation spaces in addressing the crises in biodiversity, and these suggest we should be providing more space for nature wherever it can be found and not purely in highly protected reserves.[1] Surprisingly, they reveal that urban green spaces and infrastructure can provide for an essential part, or the whole, of an invertebrate's life cycle. By providing habitats for nests or egg laying, food sources and corridors along which animals can travel, green spaces can let nature thrive. They can also provide a refuge for some species which have been driven out of the countryside by monoculture and heavily treated crops. Invertebrates in turn will give back. They will deliver essential services for the green spaces, including pollination, nutrient cycling, seed dispersal, soil managing and providing food for species up the chain, such as birds and mammals. And some joy, too. Who does not like to see a butterfly?

But how many people understand how vital that green patch is for the small, essential creatures in our lives? How many of us look closely and see what thrives when it's able to, in the grass, the flower beds or in the trees and shrubs?

In urban areas, grass verges are sprayed with herbicide to clear weeds, and the building of houses, roads and business parks covers habitats with concrete. These are all part of the decline in habitats that invertebrates need to feed in, find mates, lay eggs in and find shelter.

## Spaces for bugs

There are many spaces that, with a bit of effort and ongoing care, can be rebugged – any place where there is space for nature, however small, including the following:

- derelict land and wasteland
- public parks, squares and gardens
- cemeteries, peri-urban land (the land adjacent to an urban area)
- private and shared gardens of any size
- green verges and pavements with green patches and trees
- golf courses and horse paddocks

- sports and training grounds
- buildings with outdoor spaces – e.g. cafés, roofs and car parks – where there is an opportunity to plant some greenery or host a beehive or bug hotel

Involve your local communities and institutions using local media, resident groups and social media like local Facebook groups. If more people are allowed and encouraged to adopt a green verge or space in their street, that will create a diverse area that is attractive to worms and other bugs. Companies and institutions could support offices, retail and eating areas, and other places to become more bug friendly with a few simple actions and a bit of regular effort.

There is now, however, a growing movement to urge councils to use less harmful and more nature friendly tools to manage weeds and to reconsider even whether such control is always needed. In the past, councils were often under pressure to keep grounds, verges and parks weed free to avoid complaints about messy areas and the risks of stings or slipping on wet vegetation, for instance. But, with more public awareness, there is greater support for wilder places, and weeds that look great and encourage biodiversity. Complaints about stinging nettles or weedy mess need to be mitigated but the sight of flowers and bees flourishing on road verges and in parks, when weedkiller or mowing is reduced or removed, can help with public education. Pesticide Action Network groups across the globe are running Pesticide Free Town initiatives to show how zero chemical controls can be adopted.[2] Even the UK government's Highways Agency, which manages road maintenance, has got on board by announcing at the end of 2020 that the hundreds of miles of road verges they manage will be allowed to let wildflowers to thrive, creating vital habitat for insects and other wildlife.[3]

*Best for bees*

Bees thrive in urban areas. They may even thrive better in urban spaces than the countryside due to the diversity of plants and possibly less exposure to chemicals. The bees provide extremely important pollination services for the plants in urban areas, but they need to be able to find a good food source and to be able to locate their hives safely. One study in Philadelphia in the US, looked at the food sources used by honeybees housed in rooftop apiaries, by measuring the pollen brought back by bees and the weight of the colonies over time. They found that trees, shrubs and woody vines are among the top food sources for honeybees in that city. The bees needed different plants in different seasons. Key overall were the importance of flowering trees and the need to maintain weedy spaces, which can quickly fill with valuable flowering plants.

## In your garden

Urban gardens are a fantastic place to encourage nature and especially bugs, which will of course then attract birds and mammals. There are three guiding principles for creating a nature-friendly garden, which are:

- avoid chemicals and soil disturbance like excessive digging (and there are lots of helpful websites on maintaining a 'no dig' growing space which protects soil and bugs)

## Build a bug haven

Build some insect-friendly habitats in your home and garden, like a bee brick full of holes in or on your wall, or build a 'bug hotel' out of old hollow bamboo sticks. You can also buy one ready-made from wildlife organisations and garden centres.

Grow ivy, which provides great winter food for many pollinators when everything else is gone.

Old logs can be a haven for beetles, or dig a hole and bury some old wood for a stag beetle nest. If you are very lucky, and you may have to wait several years, a majestic stag beetle will emerge from larvae that have spent years growing fat underneath.

- don't be tidy (leave some weeds, logs and fallen leaves on the ground and the grass uncut)
- include a wide variety of plants and habitats (from flowering plants to mini ponds)

## Community gardens

You may be involved in a community garden, orchard or allotment. All can be rich in invertebrate life if managed well. Capital Growth is a London-based organisation set up to increase the amount of food grown in the city and it has hundreds of invertebrate-friendly growing sites. As urban food growers, they use many wildlife-friendly practices and principles to entice the bugs in. These include attracting wildlife into the food garden with features such as bee boxes, creating a healthy compost heap and growing nectar-rich flowering plants.[4] They suggest that counting the number of beneficial insects in a growing space to show how hospitable the area is for wildlife can also help nurture understanding of the invertebrates around.

Crucially, as with farming, what we grow in a community garden will make a huge difference. If you widen the range of food plants grown, use organic methods and leave space for wildlife refuges around the plots, you will be making a home for bugs as well as plants.

Minimising soil disturbance and using green manures – fast-growing plants grown to cover bare soil, which are then dug into the soil while still green – as fertiliser before planting will help too. Use specific plants to manage insects – like nasturtiums that lure cabbage white butterflies away from the precious cabbage plot. The whole design of a food-growing space or community garden can help attract invertebrates, such as incorporating native or non-native trees and shrubs, providing shaded areas, ponds and some grassy areas, too, for those bugs that need pasture.

## Hope for the future

I am optimistic that people, communities and even governments will start to make changes that benefit bugs. Many people are already doing so much to help invertebrates survive and thrive. Bee-friendly town initiatives are spreading fast globally and there are many communities building bug hotels and flower refuges in parks. This may be groups of individuals, schools or a community body working together to build specific areas that attract bugs.

## Good plants for insects and other invertebrates

Planting for bugs throughout the year will provide for a wider range of species and help ensure their survival over harsher months.[5]

- Spring: wallflowers, dandelion, pulmonaria, hawthorn and crab apple are great sources of early nectar and pollen.
- Summer: plants in the carrot family (such as yarrow, fennel, cow parsley) are loved by hoverflies and lacewings. Lavender, marjoram and phacelia provide nectar for deep-feeding honeybees. Buddleia, scabious and mint flowers suit the long-feeding proboscises of butterflies. Sunflowers can provide great places for weaving spiders to create webs.
- Autumn: Michaelmas daisies (asters), heather, dahlia and ivy are some of the best plants for a late supply of pollen and nectar and provide great shelter.
- Evening scents: honeysuckle, jasmine and sweet rocket are all good for moths.
- Balconies and pots: grow edible shrubs and herbs such as borage, lavender, rosemary and thyme.
- In ponds of all sizes (even a tiny pond will provide water and habitat): white water lily, yellow iris, marsh marigold, milfoil (for oxygenating the water) and frogbit are all great.

### Rebug the golf course

Other sorts of parks can be managed better for both wildlife as well as human visitors. Take the golf course: a major recreational land grab, which often excludes both human and invertebrate alike by their fences and dull manicured lawns. There are over twenty-five thousand golf courses worldwide and there are many being newly built each year.[6] That is a lot of land with the potential to be designed for nature with more habitats, which in turn reduces water runoff, irrigation, and chemical inputs.

Given their size they can be a space for invertebrates, especially with scrubby, uncultivated areas.[7] We should push for more naturally landscaped golf courses, with full public access too, especially if such courses could be built on degraded landscapes such as landfills, quarries and eroded land. Given their size, frankly, I think that all golf courses should be obliged to maximise their design to boost bug numbers and diversity.

I've seen amazing solitary bee hotels on a café wall, made with hollow tubes stuck together in which the bees lay eggs. The bee puts in some pollen food and seals the opening with cut leaves or mud to protect the emerging baby bee. Seeing those sealed ends and then the emerging bees is an incredible joy. Councils are leaving large sections of parks to be sown with wildflowers, so we get varieties, colours smells and bugs to look at rather than just cut grass.

*Urban food projects*

Incredible Edible is an inspiring initiative started by a group of friends in 2008 in a small town called Todmorden in the UK. They believed food growing could be a great way for a town or community to show it is connected, kind and able to work and learn together by growing food in often neglected or blighted urban spaces. They involve local people, schools, businesses and institutions, and have found that by growing and celebrating local food, often in strange and unexpected places, such as fire stations and roadside verges, communities are strengthened. As of 2021, there are 148 Incredible Edible groups across the UK and over 1,000 globally. They largely follow organic principles, using natural methods to manage pests and growing carefully with nature in mind. As many of these bring flowers, trees, bushes and healthy soil to places where there was none, the numbers of worms and pollinators, like hoverflies and bees will be boosted. They are also about teaching people and communities about nature – a crucial part of the rebugging toolkit.

*Urban tree growing*

Another project I admire and which is all about the bugs in the trees is The Orchard Project. This aims to get every household in the UK's town and cities within walking distance of a productive, well-cared-for, community-run orchard. It is a wonderful ambition, which will be, by default, creating more space for bugs as they use nature-friendly methods and lots of traditional fruit and nut varieties.

Traditional orchards are a priority habitat for several species of invertebrate including four hundred species of saproxylic invertebrates that live on decaying wood. These include the stag beetle, violet oil beetle, and the beautiful and rare noble chafer beetle. The Orchard Project trains volunteers and apprentices in managing community orchards for wildlife, nurturing the rebuggers of the future. They will, with the local community, create new orchards or restore ones that need help, both small and large. And a key principle is access for all and everyone can be involved – you don't need to be any sort of expert to be part of these projects.

Tiny forests are a new concept to put trees in urban areas across Europe and sounds like an idea whose time has definitely arrived. It was inspired by an initiative by the Japanese botanist Akira Miyawaki, who has created over one thousand tiny forests in Japan and elsewhere to bring nature as well as locking in carbon (always good to help tackle climate change) into the heart of the city. The first UK one was the size of a tennis court and was planted in Oxfordshire in 2020 to tackle urban wildlife loss – six hundred native trees, including oak and dogwood, filled a two hundred square metre plot.[8] It is a partnership between local volunteers, the local council and a global charity called Earthwatch to grow native woodlands in an urban setting. Many of these tiny forests have already been planted in the Netherlands and reports suggest that they attract hundreds of different animal and plant species once planted.

I am in favour of far better and well-enforced laws to stop the building of developments in key bug areas, and of better urban planning so that we have less development and roads in green areas and more rebugged urban landscapes. Another solution may be to provide effective rewards for protecting bug habitats like meadows and grass verges – to provide an incentive to developers wanting to build in an area with important wildlife, for example, by granting permission on condition that they recreate even better habitats elsewhere.

## Rebugging actions

If you have one minute:

- Let the spiders and other bugs live in your home if they want to. Keep those that you need out, such as ants, moths or spiders using harmless, non-chemical methods.
- Avoid using pesticides in the home, such as for moth treatment or ant removal – use alternative methods and work to stop food pests gaining entry through gaps, doors, cracks and windows, and by ensuring food is not accessible to them by storing it in closed containers.
- Work around holes through which these bugs can enter with soapy water, cinnamon, lemon or pepper spray. Planting mint can also be a good deterrent. If you have to kill some, use boiling water like my mother or cornstarch (which smothers them), not chemicals.
- Moth infestations can be hard but there are lots of tips online for getting rid of them including freezing clothes and using moth repellents such as cedar wood and lavender.
- Never step on a bug, and leave some aphids and blackfly alone. Ladybirds, lacewings and birds will feed on them and provide an ongoing, self-replicating, pest management service, free of charge.

- Take some time out to see the bugs. To spot insects the best thing to do is nothing! Just sit still and they will come.
- Leave dandelions and other native wildflowers as feed for bugs. The types of invertebrates that visit will depend on many factors but the messier your garden, the more flowering plants and hiding places, the less chemicals and interference (such as excessive mowing and weeding), the better.
- Be untidy in your garden, window box, allotment or balcony. Insects love weeds. Other invertebrates love dead and decaying wood and leaves. This may even save you time.

If you have more time:

- If you have just a balcony, porch or window ledge, grow and nurture some flowers for bugs, like catmint, geraniums, sweet alyssum and lavender. You will be providing a vital food source or shelter as the bugs travel through or by your home.
- Garden without pesticides – great tips are available from Garden Organic and other organisations. If you have an allotment or community garden, you have a huge opportunity to rebug by going organic and adding wild areas – encouraging beneficial insects

and other invertebrates onto the plot to tackle pests and weeds.

- If you have some lawn, keep the grass high as this provides shelter for invertebrates and, if you can, let some or all of the grass and weeds flower as that will provide a great food source for many bugs.
- Grow your own vegetables and fruit, without chemicals. Many will be flowering plants that provide food and habitat for insects.
- If weeds really are growing in the wrong places, use brute force, boiling water or vinegar to remove them.
- Having an open compost heap will provide habitat, warmth and food for species like beetles, ants, centipedes, snails, earthworms and millipedes.
- As you cultivate your garden or grow your pots, try to keep some seeds for next year. Using your own saved seeds or organic certified seeds – produced without chemical pesticides or fertilisers – will help reduce chemicals in the environment. Look for local seed-swapping events or start your own.
- Don't release imported animals and plants you have bought from pet shops or garden centres into the wild as they could harm native species, and don't import non-native

species yourself. Ideally buy plants only from local nurseries who grow from seed.

- Build up soil health in your garden (many invertebrates like worms and many insects live in it) with home compost and homemade or organic feed. Avoid too much bare soil, which may be eroded in heavy rain or wind, and too much digging, which also disturbs the delicate systems below ground.
- Visit a rewilding project, like the Knepp estate or Cairngorms National Park. There is often the opportunity to hire accommodation on their grounds – like at Knepp and Cairngorms – so you can stay to spot some of the invertebrates that are now thriving there. At Knepp, that includes the rare purple emperor butterfly and the violet dor dung beetle. Over six hundred invertebrate species have been recorded there.

If you can go that bit further:

- Plant wildflowers whenever you can at home, work and local green spaces – they are a great food and brood source for invertebrates. Prioritise nectar-rich flowers, herbs and grasses, too, and choose native, locally sourced and grown trees and plants. Avoid double-flowered cultivars – these often have less or no pollen and nectar. Aim to make sure there is something in flower all year round.

- Plant some trees, hedges and shrubs. These are hugely important for invertebrates as they provide food, shelter, habitats and corridors for them to move about safely. If you are not able to plant some yourself, get involved in local tree planting initiatives. Many councils are now planning to plant trees to counter the climate and nature emergency and will have tree officers you can contact. Join your local park friends group.
- Avoid having to buy compost by building up your own compost in a compost heap and liquid fertiliser by using a wormery. You can buy tiger worm eggs online and they are fantastic at converting your food waste into valuable liquid plant food and rich compost.
- Make a pond – it does not have to be big but it can provide valuable habitat for some bugs and a water source for many others. Many invertebrates rely on these freshwater habitats as a permanent home, food source or a place to breed.
- Join community groups or clubs in your area dedicated to improving habitats for invertebrates, like wildlife groups or orchard projects.

— CHAPTER 5 —

# The bigger bug challenges

From the bees, ants and worms to wolf spiders and springtails, we know that keystone species – those on which whole ecosystems can depend – are at risk. Creating amazing spaces, small and large, will do much to help and could even mean the difference between a species going extinct or not. But there are major forces at play that will need greater collective action if we are to reverse all the harm. Our activities for so long have had such a huge impact – on the climate, on seas and on nature – and we have polluted the water, air and soil too much. To stop the deep and complex drivers of invertebrate decline we will need international and national, as well as personal, action. There are also huge vested interests involved. Can we respond with the right measures and actions?

Butterflies, bees, ants and dung beetles seem to be the insect species known to be most affected in the UK (although that may reflect the focus of most research). We have also lost four water-living insect groups, including some dragonflies and caddisflies. You may expect specialist species – those who can only live in a narrow habitat or need a specific food supply – to be the most vulnerable as they cannot adapt quickly, whereas species that are able to adjust their habits quickly can fill gaps left by others. Specialists are being lost but, worryingly, analyses are showing that generalist, common, species are also in decline.

Overall, the total mass of insects is falling by a highly disturbing 2.5 per cent a year, according to the best data available. But we need to know more because the data is still incomplete and too restricted in scope.[1] One major problem is that some studies take local species declines and multiply this to regional or global levels, which may not show the true nature of change. Each field, let alone country, will be hugely different so big extrapolations can be misleading. It really is challenging to assess invertebrate losses and only a small fraction of insects has been monitored long term. Many (though not all) of the studies so far inevitably have come from managed landscapes in western and northern Europe where there is money to pay for it.

All these suggest to me that we have a problem but we really know too little about it.

## The impact of climate change

Human-caused climate change is an existential threat to all of us and not least in how it will affect the invertebrates we rely on for so much. I have been working in the environmental movement for thirty years and in that time, climate change has grown into the greatest crisis we are all directly or indirectly trying to tackle.

Being small, all invertebrates are highly vulnerable to changes in temperatures, weather extremes, droughts and rainfall patterns. They have a large surface area to volume ratio so, despite having lots of tricks to avoid it like waxy skin, they are susceptible to drying out. Their growth is also highly dependent on temperatures. A forty-year study of flies caught in suction traps in the UK's Rothamsted Research, showed

**The amazing monarch butterfly**

The fate of the stunning monarch butterfly in America is gaining significant attention. This insect migrates as flying adults in astounding numbers up the continent from Florida and Mexico to overwinter in western states of the North American continent. When they migrate, the monarchs cover thousands of miles, breeding new generations on the go. Over the last few decades, their overwintering numbers have plummeted to less than 1 per cent of the population size in the 1980s – a critically low level. Climate change, deforestation and habitat loss are all possibly to blame. Pesticides, too. A recent study of pesticide contamination of the monarch's main food source in California, the milkweed, found sixty-four different pesticide residues.[2] When we know only a handful have been tested for toxicity to this butterfly, it is clear we are not taking care of the bugs.

that the timing of peak flight by the flies has advanced by an average of seventeen days, from 23 July in 1974 to 6 July in 2014. In addition, one third fewer flies were observed.[3] These changes will have a knock-on effect on birds, as flies are an important food resource and need to be around in sufficient numbers at the right time. There is often a close correlation between bird declines and the decline of insects. Such worrying changes in life cycle timings and numbers have been found in many other studies and climate change is likely to be a major factor.

Another example of where we may be affecting something profound is with the tiny marine bugs, called zooplankton, which include microscopic species like krill, which look like tiny prawns, found in vast numbers in large bodies of water. They feed on plant plankton called phytoplankton and control its growth, making up a huge part of the base of the food chain.[4] Plankton play a central role in ecosystems, nutrient cycling, ocean currents and fisheries, yet there has been a decline in eight out of ten regions recently studied.[5] Rising sea surface temperatures due to climate change are thought to be the main issue. And there may be an added scare as the phytoplankton absorb greenhouse gasses. If we upset the plankton balance, we affect their ability to draw down this carbon dioxide. A frightening feedback loop.

It is now known that bumblebees, crucial for flowering plants the world over, are affected by the increasing number of days with extreme heat. Greater local extinction rates in bumblebees have been seen across North America and Europe, resulting from an increasing frequency of unusually hot days. This means fewer bee colonies, so fewer species within a region, which will lead to wider impacts on plants and other species dependent on them. As we expect average temperatures will continue to rise, meaning more extreme days too, we may see the hugely iconic, fluffy but also critically important bumblebee unable to survive.[6]

It is also not just about insects. I often get frustrated at the headlines focusing solely on insects when there are significant concerns about other, crucial species such as worms and spiders.

*Earthworms: expert soil engineers*

As far as I am concerned, there is little not to like about earthworms. They are truly a keystone species helping to hold the ecological systems in which they live together and having an impact way beyond their size. With around seven thousand species, they are hugely diverse and in a hectare of good soil you would expect to find around four million worms busy borrowing, eating and multiplying.

Earthworms ensure nutrients are available in the soil, either directly or by encouraging microbes, which act together with plants to make nutrients available. This can increase crop yields by 25 per cent. They create soil structure that allows moisture to be retained and boost carbon storage in soils as they digest and release plant residues. I cannot imagine humans being able to recreate that hugely complex role with a 'robo-worm'...

Sadly, a loss of earthworm populations in agricultural soils has been documented in many countries. In response to concerns, a recent global initiative by 141 researchers from 35 countries created a global-scale atlas of earthworms mapping global diversity, abundance and estimated weight of the global population.[7] The project revealed how temperature and rainfall can affect earthworms, so climate change is likely to have serious implications for both earthworms and the services they provide to nature. Worms are also semi-aquatic, needing sufficient moisture to survive. They looked at likely culprits for the worms' decline and found the impact of intensive farming practices disturbing soil and chemical use was key.

Dung beetles are another keystone species, which we need to create good soil structures and to process animal dung, and they too are sensitive to temperatures. These vital bugs have been found to abandon their dung if temperatures are wrong and their breeding is affected.[8] A disaster waiting to happen if we found our livestock fields without this vital soil improver and nutrient recycler.

*Spiders getting hot under the collar*

Spiders are another species I feel we should appreciate more. Yes, they have fangs and eight legs, can be hairy and a few species are dangerous. But they are extraordinary. With over forty-eight thousand known species, they have been able to adapt to almost all habitats on earth and even sort of fly using electricity (see 'Electric leaps' in chapter 1, page 31). They play a crucial role in food webs, both as predator and prey, and undertake a fair amount of plant pollination. They also spin silk.

Some studies have suggested climate change may cause spiders to become bigger and more aggressive.[10] The aggressivity probably emerged as the studies were made on a group of spiders – social ones called cobweb spiders – living in tropical cyclone areas. When species live in areas with frequent extreme events like cyclones, they are more likely to survive if they have more aggressive traits compared to more docile ones. This is possibly because they need to bully others when having to recolonise an area quickly after the devastation. The researchers in this study had to drive straight into the storms to do their research – an impressive commitment to spider study. As extreme weather events become more frequent with climate change, their conclusion is that spiders may become more aggressive over time as selection pressure means more aggressive spiders will be chosen by females more often.

A separate study of the beautifully named wolf spider in the Arctic tundra showed that earlier springs and warmer summers caused by climate change may lead to larger female spiders and more offspring. Warmer temperatures mean more prey, which leads to greater spider population densities.

**Skilled weavers**

Researchers investigating a specific spider species that creates complex webs found a unique comb structure on their legs that stops the silk fibres from sticking together. When they shaved the spiders' legs, the fibres tangled. Artificially produced nanofibres so far have been tricky to handle in the laboratory as they tend to stick to the equipment. The spider study showed how combs could be used to keep fibres non-sticky, providing valuable insight into potentially useful new technologies.[9]

They eat more prey but they also resort to cannibalism and eat their young. These spiders are a key invertebrate predator across the Arctic, so changes to their numbers and activities could seriously affect the whole food web. A particular favourite prey is the fungus-eating bug called the springtail. If wolf spiders eat more springtails, you may think this would increase fungal growth and so speed up fungal decomposition of plant litter, so speed up greenhouse gas release – a major ecological impact. But, in a rather clever study, researchers found the opposite was actually true. More spiders resulted in fewer springtails eaten – and so more springtails eating fungus, so less fungal activity and less decomposition.[11] One suggestion for this was more cannibalism of young spiders by the larger females, instead of eating normal prey, leading to more springtails eating more fungus. It is a beautifully complex web that we are only beginning to understand. And we disturb it at our peril.

## The impact of pollution

We have created a toxic soup in rivers and seas, in the soil and in the air. Alongside climate change caused by greenhouse gasses, invertebrates are harmed by so many pollutants from the chemicals used in agriculture and in homes, from herbicides and insecticides to noxious air pollution from cars and livestock systems, sewage waste and nutrient pollution in water.

Pollution can kill bugs directly or it can have chronic effects such as reducing their ability to navigate or to breed, and also hurt by harming the plants or prey bugs rely on. The widespread use of weedkillers in America, for instance, has removed the crucial milkweed that monarch butterflies use to lay their eggs.

When looking at pollution as one of the factors harming diversity, a top-level international panel of experts found that fertilisers leaching into coastal ecosystems have created over 400 ocean 'dead zones', totalling more than 245,000 square kilometres – an area greater than the United Kingdom. They also calculated that plastic pollution has increased tenfold since 1980 and that 300–400 million tons of heavy metals, solvents, toxic sludge and other wastes from industrial facilities are dumped annually into water habitats.[12] All this creates catastrophic conditions for most invertebrates and only the most hardy will survive and thrive.

Looking at the UK, the Environment Agency collates pollution incident data and recent assessments suggest serious or significant pollution incidents have been rising, though it does vary from year to year. Agricultural pollution and run-off are the major issues, but water companies,

## Key actions to combat climate change

- Governments need to act to reduce greenhouse gas emissions from industry, transport, heating, land use and so on, but they must work internationally as this problem respects no borders.
- Land use changes to combat climate change, such as tree planting to store carbon or new energy crops to replace fossil fuels, must do no or minimal harm to invertebrate populations and habitats. A commonplace phrase sums this up well: 'the right tree in the right place' so that we are not harming, say, a vital bug habitat like peatland to plant trees or a bug-rich rainforest in Malaysia to grow palm oil as an alternative to petrol!
- Farming must play a role in reducing emissions and capturing carbon in soils and trees.
- Citizens, as homeowners and as consumers, need to get the right kind of support and price signals to reduce emissions and decarbonise their lives, homes and shopping.

too, have been responsible for a considerable number of acute sewage pollution incidents that harm aquatic invertebrate species.

### Masterful climate control

The ability of tiny invertebrates to use social homeostasis is a fantastic example of living together efficiently. Social insect colonies, such as termites, work together to sustain the colony environment – its temperature and humidity – in the same way that cells of a human body coordinate their activities to keep the body's internal environment at steady levels. All that takes energy, so the organism wants to use the most efficient way possible to achieve this. Different termites do different jobs, while the construction of their termite mound is so sophisticated it enables the environment to be perfectly controlled.

As temperatures rise under climate change, learning from these bugs is invaluable. They have survived for millennia without making a mess of their surroundings. And we are not talking small either. In Brazil, scientists have found a vast area of 200 million termite mounds spaced on average about 60 feet apart.[13] They have found that in this still-inhabited ancient termite city, as large as Britain at 230,000 square kilometres, there are some mounds that are older than 4,000 years. But it is all *one* colony living in a huge, underground network. A colony, working together, creating steady environments in homes built to last.

Scientists at the University of Nottingham in England have studied such systems with a view to seeing how we could design our buildings better.[14] The gas exchange and ventilation termites use in a complex set of internal galleries, tunnels and duct could be mimicked in human structures to provide better systems than our energy guzzling air-conditioning systems for instance.

Pesticide use globally has been increasing and is responsible for significant invertebrate decline both directly in killing wild populations in order to destroy crop pests, but also indirectly by removing prey species, plant habitats and food sources. According to the country data collated by the Food and Agriculture Organization of the United Nations, use of pesticides in kilogrammes per hectare has increased in all regions since 1990. Pesticide residues contaminate soils, water and marine environments, and the chronic and synergistic effects, where chemicals work together creating unexpected effects, are only just beginning to be understood.

The pesticide treadmill is hard to get off and this is shown clearly in the build-up of resistance to the pesticides in target species. When one chemical fails because the pest or weed evolves to

be resistant, the solution is to find a new one and so the cycle continues. And this could have catastrophic effects for public health: mosquitos, vectors of deadly malaria, are becoming resistant to the main insecticides used to control them in Africa.[15] Over-reliance of pesticides in farming is also generating farm pests and weeds resistant to the usual controls – they are literally evolving ways to be resistant to the pesticide and so can no longer be controlled. We may find ourselves facing food gaps or serious food shortages unless we change our methods of food and fibre production.

## Invasions of the wrong bugs

You may have heard of the 'murder hornet' or the Asian giant hornet that was found for the first time in the US in 2020. This is the world's largest hornet and is a scary-looking beast, but it does not deserve the 'murder' nickname. In the UK we have the smaller Asian hornet, which was accidentally introduced in 2004 from China that is now spreading all over Europe and causing problems. Both are great insects but they are just in the wrong place and, as a highly effective insect predator, the hornet can cause significant losses to our bee colonies and other native invertebrates. They sit outside honeybee hives and grab workers as they go in and out, chopping them up and feeding their juicy thorax to their young hornets in their nest. Bees can't sting them but some have now learned to heat them to death (more on this later).

Invasive species and diseases are identified as a top threat to wildlife. They can upset the delicate balance of a food chain and can be hugely destructive. In the UK, there are more than 3,700 non-native animal and plant species. As global trade and human movements increase, more invasive species have been identified because moving people or goods gives invasive

## The Namib desert beetle and water conservation

Water is obviously a basic need. Finding new and low-energy ways to conserve water is becoming a considerable task as we face droughts and more irregular weather patterns due to climate breakdown. An invertebrate providing some useful insight is a species of darkling beetle, the Namib beetle, which lives in the African Namib desert, one of the driest places on the planet. It is a tiny beetle, but it has evolved an incredible behavioural and physiological way to pull water out of air.[16]

On cold mornings, when a fog wafts in from the Atlantic Ocean, the insect makes its way to the top of a dune and does a head stand, tilting its back into the breeze – a stance called fog-basking behaviour. Its waxy exoskeleton, with a pattern of nodes along the beetle's back, create a large and purposefully designed surface area that collects moisture from the morning fog. The droplets are then directed into small channels and on towards the beetle's mouth. Academics at Massachusetts Institute of Technology in the US have studied this design and are currently developing new water-catching materials that will be cheap, durable and more effective than current approaches.

species a hitchhike into new habitats. Records of alien species have increased by 50 per cent in the last 50 years.

But it is not always possible to stop invasions. The colourful harlequin ladybird or ladybug, which comes in several colours, eats our native ladybird species as well as their prey. We know how useful our native ladybirds are for controlling aphids and other pests, so an invasive insect that eats their babies is unwelcome. But controlling them is tricky as they have settled in everywhere – they are now too numerous to destroy individually and any chemicals would harm our native beetles. We may have to just let a new balance of beetles take shape.

Ultimately, we need greater controls on the trade in goods that can bring in invasive species. And we need strong public campaigns so that everyone is aware of the dangers. Support for more domestic plant nurseries would also help so they can safely provide the plants people need – ideally native species grown without chemicals.

## An artificial environment: light and noise pollution and wi-fi/5G

I was intrigued and alarmed to find out that some of the ways we run our lives now creates new dangers for the bugs. We need a far better understanding of these potential problems, particularly as they are all growing in intensity.

**Too much light?** The increase in the distribution and intensity of artificial light can disrupt natural invertebrate behaviours and life cycles.[17] Moths, which are crucial pollinators, are attracted to artificial lights at night, and aquatic insects can be attracted away from water by light shining on surfaces. When bugs are exposed to unnatural light patterns, such as all-night street lights, it can disrupt the natural light-dark rhythms they use for regulating activities and for movement. This can disrupt their feeding, breeding and movement, leading to lower breeding numbers and more isolated populations as they are not mingling as they would normally. International studies have confirmed the negative impact of artificial light on invertebrate species, from snails to aphids, and confirm the need for action – for instance, to specify municipal lighting design that involves better timers and minimises glare and illumination, such as narrow-spectrum LED lighting, which is less attractive to most insects.

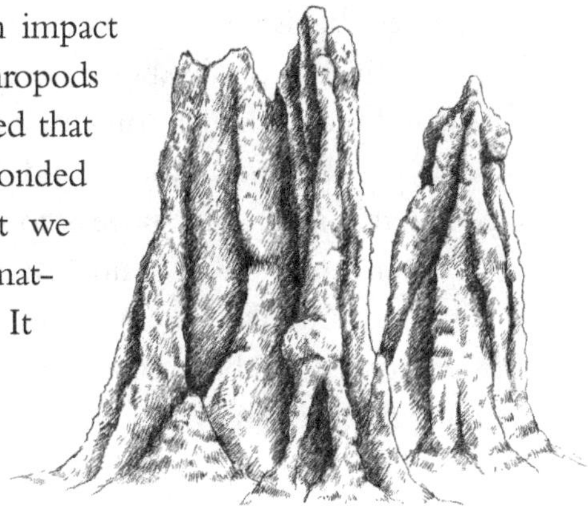

**Too much noise?** One study on noise pollution found that the hermit crab selection of which shell to inhabit was significantly affected by noise.[18] When exposed to noise, the crabs approached the shell faster, spent less time investigating it, and entered it sooner. We can imagine how this may well affect the crab's survival if it ends up choosing a low-quality home. Noise is a surprisingly serious environmental pollutant, now recognised as a major global problem. Bugs can detect noise but they also use noise to communicate and it is only recently that scientists have begun to see how we are harming invertebrates by the noise we make. A 2019 review of over 100 studies on noise pollution impact (including on arthropods and molluscs) noted that most species responded to noise, and that we may be underestimating its impact.[19] It can affect development, bodily functions, and behaviour. Crickets, bush crickets and grasshoppers are extremely sensitive to sound as they use it to communicate over large distances. In one study of a spider, which uses vibration in courtship to communicate with potential mates, researchers found that its mating success was reduced by white noise of 0–4 kHz (the top end sounding like a very high-pitched whistle). We need more research, but it is clear we may have a significant and overlooked pollution factor affecting invertebrates. Should we all be less noisy if we want to rebug?

**Are our smartphones cooking insects?** Most of us will be totally unaware that our phone and wireless signals may be causing harm to invertebrates when they are exposed to the electromagnetic fields (EMF) they create. These fields are basically invisible sources of energy or radiation and can be absorbed and cause heating in bugs. Wireless signals, such for our smartphones, are now the primary outside source of these fields and the impact may be accelerating as we move from older, lower frequency phones (2G, 3G, 4G phone signals and wi-fi) to higher ones (such as 5G). The energy involved in these newer systems is higher, so increases the heating effect on any little body, such as a bee. In one of the first studies to investigate this issue, researchers looked at the absorbed power in the dor beetle, the western honeybee, the desert locust and the Australian stingless bee.[20] Thankfully using models, not real insects, they found that frequencies above 6 GHz – that is, 5G and above – could heat the insects up to 370 per cent more than normal levels. What about the much-loved but also economically important pollinator, the western honeybees, who are continuously exposed to these frequencies as they fly? Using modelled impacts on workers, drones, larvae and queens, but also using

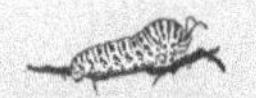

## Bug wi-fi

Many invertebrates have already formed their own 'wi-fi'. Super organisms of many millions of individuals can communicate using a wide variety of tools including chemicals called pheromones, sounds, touch and visual stimuli; for instance, the honeybee's foraging 'round dance', which encourages other bees in the hive to go to nearby nectar sources and their 'waggle dance', which specifically tells bees the distance and location of a good source of nectar it has found further away. They are also expert at pheromone use. The brood (larvae) will release chemicals to stimulate more pollen foraging and adult bees release pheromones to alert the hive to dangers. I worked one summer at an experimental station testing what chemicals and cues honeybees used to stimulate foraging or fight behaviour by other bees in the hive if there is a risk. I spent many happy hours sitting next to beehives in the sun, counting bees coming in and out. Ants, too, use all these tools extremely effectively to manage complex communication requirements. Each chemical or other cue is designed to tell a specific worker what to do so that they are clear in their role. This is often touted now by management gurus as a great example of teamwork management with the top learning that clarity of roles is key.

real exposure measurements near beehives in Belgium, researchers estimated realistic exposure levels.[21] They found that even a small shift in power to higher frequencies will lead to an increase in absorbed power by bees by between 390–570 per cent. This could lead to changes in insect behaviour, bodily processes and shape over time due to an increase in body temperatures. Again, more research is needed, but with the wider roll out of 5G that uses a much higher frequency, we could be creating an invisible killer issue for our bees and other bugs.

**Rebugging actions**

If you can do one thing a week or month:

- Cut your use of harmful chemicals that end up going down the drain or polluting the garden.
- Ask your vet for non-neonicotinoid pet treatments for your pets and use physical ways to reduce fleas or tick such as regular grooming, washing and combing.
- For cleaning, use biodegradable products or use the old-fashioned, but cheap and effective alternatives such as vinegar and baking soda. And avoid putting any strong chemicals down the drain where they can reach water sources and harm invertebrates.
- Choose low-level lighting – buy low intensity lighting and only put it on when needed.

- Leave dark spaces in your garden and position lights as low as possible. Solar lighting is cheap and produces an ideal, soft light.
- Avoid a 5G phone if you can and, if not, use it only in confined areas when possible.

If you can go a bit further, become a bug campaigner:

- Reduce as much as possible climate-change emissions in your life – cut flying and driving, and instead walk, cycle and use public transport.
- Make your house energy efficient, eat a sustainable diet, reduce food waste, reuse and recycle. This helps all life, including invertebrates.
- Join a great bug champion organisation. There are many suggestions in chapter 9. And if you are doing conservation work locally, make sure everyone there understands we need political action, too.
- Join and take actions with organisations that are running climate campaigns as runaway greenhouse gas emissions are one of the biggest threats to bugs and their habitats. See 'How to start lobbying for change' in chapter 9, page 172.
- Ask your local council to provide organic food and sustainable menus in public canteens, schools and universities, and to use bug-friendly methods on their grounds and lawns.

- Ask local schools how they are promoting invertebrate care on their grounds and teaching environmental issues to students.
- Use the media if your council does not respond. Write to the local paper, join a local group, use social media to encourage them to act with gentle or more firm persuasion. Parliamentarians and councillors are keen on being seen to support nature and I have found they often want to be photographed beside a giant furry bee.
- Support organic campaigns which help farmers and growers to use nature-friendly pest management.
- Pick a key issue you feel strongly about – such as a company you feel should be doing things differently locally or globally, or the protection of a local tree or nature area – and focus on that. Become a lead campaigner and draw in others to help you. Start a new group if you need to.
- Ask your family and friends to pledge to take action too, and always share what you are doing on Facebook, Twitter and Instagram.

— CHAPTER 6 —

# Why our farming, food and shopping need bugs

We are going nuts for almonds these days. But the dark reality is huge, thirsty, nature-free, almond tree plantations, needing millions of bees brought in to pollinate the crops. Around 80 per cent of the world's almonds are sourced in California's Central Valley, where bee farmers truck thousands of colonies of European honeybees, categorised as 'livestock', to huge plantations. They routinely lose 30 per cent of their 'livestock' as the bees won't survive in such a hostile, sterile place. Wild pollinators like native bees and hoverflies also cannot survive in those uniform rows of trees with no other habitat around.[1] This is proving catastrophic to local wildlife.

We know farming can crash invertebrate numbers by harming habitats, but one study on Canadian farmland looking at the number of butterflies, hoverflies, bees and ground beetle species found one of the biggest factors for their presence and abundance was field size.[2] Cultivation, pesticides and crop diversity were important, too, but the field size impact shows how important 'green corridors' are for invertebrates to travel through. The edge of the field

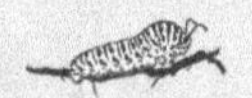

**Clever farming bees**

As if they were not impressive enough already, bumblebees have been found to induce plants to flower. Buff-tailed bumblebees make tiny incisions in the leaves of greenhouse plants but don't consume the bits of leaves or use them in nest construction. Researchers tested the idea that the bees were telling the flower something by creating a low pollen situation and showed that, yes, the bumblebees can force plants to bloom up to a month earlier than usual, a useful skill to use probably at times when nutrients are scarce. If bees can do it, maybe farmers could, too, induce flowering in certain situations. Though whether we should, is another question.[3]

boundaries, with messy bits, trees, hedges and so on provided a vital 'motorway' as well as being habitats for invertebrates.

As with any animal, when we remove or change the homes where bugs live, feed or breed, they will not thrive. If we take away the green 'corridor' such as a wood, roadside verge or hedge that they use to colonise new areas, their offspring will be stuck and will not thrive. Sadly, bug homes are being lost at a rapid rate globally. Land conversions to monocultures of crops and trees have replaced rich grasslands, forests and wetlands, alongside huge urban sprawl in many parts of the world.

**We should talk about the apple snail**

Why is it so important? Well, it has a snorkel and it lives in the unique Pantanal wetlands covering parts of Brazil, Bolivia and Paraguay, which is home to many unique species including this extremely handy freshwater snail. The natural annual flooding of its wetland habitat causes the plants to die and start to decay. This means the water becomes deprived of oxygen as the microbes decomposing the plant matter use it up. Most other animals normally involved in consuming decaying plant matter won't survive but the apple snails have evolved both gills and lungs just for this purpose. They have a long 'snorkel' which they extend out of the water, taking air into their lungs so they can live in the deoxygenated water. So, they can continue to eat through the plants and recycle the nutrients in them, which then becomes food for other plants in the area. They also have a flap to seal their shell entrance to prevent drying out when buried in mud during dry periods, and, finally, they also provide a juicy morsel for many animals. So, overall, a keystone champion of the Pantanal.

But forest and scrub clearance for cattle and soy farming is destroying the delicate river and land systems in this area, as it is in many parts of the world. Invertebrates, including insects, represent a large share of species also found in the Amazon in South America – and more are being discovered all the time. Lemon ants, heavily armoured rhinoceros beetles and stunning blue morpho butterflies to name a few. Some scientists estimate that 30 per cent of the animal biomass of the Amazon Basin is made up of ants. But they are sadly being wiped out by land cleared for grazing or for feed crops, like soya.

You will not find a great variety of invertebrates in monocultures – they are intensively managed to remove both plants that are not crops and unwanted bugs to reduce disease and pests. One recent study looking at wild pollinators in Britain found evidence of declines across a large proportion of pollinator species between 1980 and 2013.[4] This coincides with a huge change in farming as complex landscapes were made uniform and specialised. Fields were merged, so field edges, trees and hedgerows were lost, more chemicals were used, pasture meadows replaced by fertilised grassland and mixed farming became less profitable. All these created less favourable conditions for most bugs and especially pollinators.

All this uniformity is desired and dictated by the modern food industry. To rebug through better habitat care, we will need to increase the knowledge and capacity of those managing the land as well as ensuring food markets support them, through a shift to more invertebrate-friendly farming methods.

## Land sparing and sharing for bugs

The idea that we should encourage farmers to shrink the size of fields they have spent many decades increasing – to cope with larger, more efficient machinery and produce more food – is quite a hard one to contemplate. But it could have a huge impact.

At the very least we should be protecting the smaller fields and boundaries like hedges that do still exist and rewarding farmers for nature-friendly farming such as those who apply organic approaches. And there is some hope, too, as more innovative, smaller and lower-impact machinery, such as small robots and drone technology, are being developed that could create a real opportunity to change field sizes or at least maintain those small fields that are left.

Like the field edges, there is a strong case to make for policies on 'land sparing' or creating uncultivated reserves to protect the remaining pristine, undeveloped areas and to restore or to recreate important sites, maybe through deep rewilding. This could be both whole areas, like the US wilderness areas, or leaving uncultivated bits in connected networks to ensure wildlife can find habitats and move through safe areas.

Such 'sparing' of land for wildlife is used in many areas where reserves such as wildlife sanctuaries and parks are created for key species. I've not often seen invertebrates as the 'chosen' species – more often it's the larger mammals like wolves or rare plants that's the focus. So, it's great to find a UK project that is protecting threatened bugs. Buglife is working to bring 'Back from the Brink' invertebrates, such as the spectacular-looking but also catastrophically rare ladybird spider, by protecting low-land heathland sites in Dorset. The species was thought extinct in the UK, but some were rediscovered in 1980 in Dorset and with much work in protected areas, there are now fourteen breeding populations and more protected areas planned.[5]

Where land is in use, we need to creatively support farmers in the nature *sharing* approach because one thing is clear – without public intervention, farmers will have an almost impossible task balancing the demands of wildlife and the need to produce food crops and livestock at ever lower cost, as well as more rapidly and more uniformly. We need to change consumption patterns, prices and cosmetic expectations to tackle this huge issue.

This chapter explores how our choices of what we buy can drive more invertebrates back onto the land and into our lives. It is time to look at our everyday habits – our foods and clothes – and to look again. And none of these actions or changes of habit will stop you getting your kids fed or to school. Unless you get really activist and glue yourself to a polluters' van…

## Half a caterpillar in your apple is better than none

The old joke 'what's worse than finding a caterpillar in your apple – half a caterpillar as you've eaten the rest' needs a new spin. Seeing the odd bit of evidence of nature in your food – a caterpillar on your apple or aphid in your lettuce – is actually desirable because the alternative is a heavily sprayed crop, bred and cultivated to remove most signs of life.

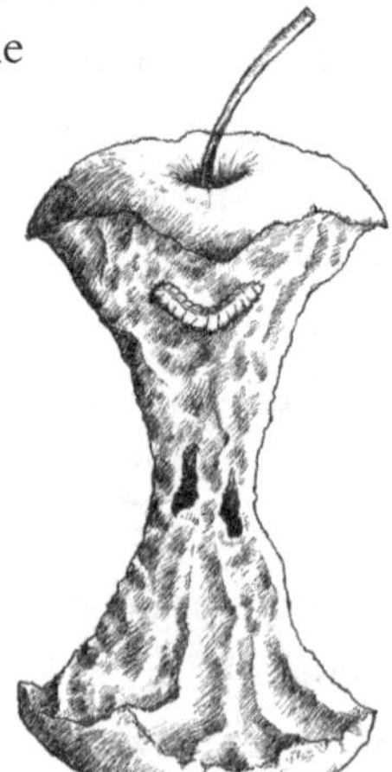

A different attitude to bugs and imperfect produce would also help the farmer to grow their crops with rebugging in mind. We have been led to expect carrots of uniform length, colour and shape. Apples all the same size, with no blemishes or bumps. To achieve this, however, growers have to use identical varieties and chemicals to protect them and make them look 'perfect'. And often they have to throw away or sell at much lower cost those that don't make the 'grade'.

We know that many factors are involved in the decline of bee populations, but neonicotinoid insecticides, or 'neonics' as they are called, and which are specific to food production, have been a major focus over the past few years for environmentalists.[6] The chemicals change the normal activities of the bees – such as when they forage for food, as well as their orientation and navigation abilities. It even disrupts their sleep.[7] Soil-dwelling black garden ants, like the ones I used to farm in my ice cream tub, are also affected. They have fewer workers and larvae when exposed to neonics.[8]

Because the harm to critical species like bees had been revealed, many neonic insecticides were in fact banned in Europe for field crops, meaning invertebrates have some space to recover and farmers are developing alternative approaches. There is considerable pressure to remove or weaken this ban as farmers struggle to find good alternatives quickly. But, given its harmful impact, politicians need to listen to the science and maintain the ban while simultaneously supporting the research and uptake of alternative tools like biocontrols, more robust crop breeds and rotations to break up pest and disease cycles.

But it is not just about the chemicals. There are many other ways in which a new diet could help the bugs. Sadly, the norms that have developed in modern, western food production have created a crisis for wildlife and bugs as well as our health. If the western world changed what they ate they would, with one small shift, make life a lot easier for invertebrates. I suggest rebugging provides a great incentive to make that journey.

### The naming and shaming of a crop pest

The cockchafer is one of my favourite large bugs with beautiful antennae and a bumbling manner. It is named for its large size and the German word for gnawing beetle 'kafer', and it also has a large set of common names reflecting its status as a major crop pest in past centuries. These include maybug, billy witch, spang beetle, kittywitch, mitchamador, bracken clock, bummler, chovy, cob-worm, dorrs, dumbledarey, dumbledore, humbuz, doodlebug, June bug, may-bittle, midsummer dor, oak-wib, rookworm, snartlegog and tom beedel.

The cockchafer species was actually taken to court in Avignon, France, in 1320, following huge crop damage (by adults and the larval grubs), where it was judged guilty and ordered to leave the town or be outlawed. All cockchafers who failed to comply were collected and killed.

## Rebugging our diets

Moving on from looking at *how* food is being produced to *what* food is grown is also worth exploring. This is often neglected but, as we have seen, the range of crops and livestock, and the genetic diversity of the seeds and breeds have a huge impact on invertebrate habitat and food sources. This really matters for a huge range of bugs.

Our food system relies on monoculture crops producing uniform harvests in huge quantities, cheaply and with the least disruption. It relies on huge amounts of cheap meat from highly industrial or intensive livestock systems, with vast numbers of uniform animals bred to tight specifications, needing high-protein feeds, and protected with antibiotics and insecticides.

The astounding-looking variety of foods in the modern supermarkets obscures the fact that much of it comes from a very few varieties of plants. The majority of our nutrients come from just corn, rice, wheat, and root crops like potatoes and cassava. Corn, grown in vast quantities in the US, is one of the biggest and most destructive crops – US corn production is a massive monoculture, mostly genetically modified to be tolerant to pesticides, and it is responsible for soil erosion and pollution. It is also highly subsidised, and the related corn feeds and high-fructose corn syrup industry is one of the most powerful agrifood industry's lobbyists, working to protect its subsidies, market and trade benefits.

*A case of two farms*

Imagine two farms. The first one is using many rotations so that there are at least seven different types of food types from the soil over a five-year period. Cover crops are sown to ensure the soil is rarely bare and exposed to the elements. The small fields are surrounded with thick hedges, interspersed with trees and flowers, with strips beside each hedge where the land is not cropped or sprayed with pesticides. The farmer rarely uses pesticides or fertilisers as nature does the hard work.

The farm has livestock: a native beef cattle herd grazes on species-rich permanent pasture, and a small sheep herd also

**An adaptable diet**

Food gathering is clearly an art form in many invertebrates. They can grab, pierce, suck, scrape, filter, cut, mash, store food and paralyse prey. They can be herbivorous, carnivorous, omnivorous, fungivorous and even eat nothing for months or even years. They can farm fungi, plants and other animals, and can change their diet completely several times through their life cycle. They will even provide their growing young with fresh live meat after leaving them, so ensuring their survival.

helps keep down the weeds in the fields and eats up fodder crops in the rotation. These animals deliver nutrients which invertebrates like dung beetles help to incorporate into the soil. There are multiple crops and meats produced for local processing and for sale in a variety of outlets. Luckily, this farmer has local customers who recognise the value of what they are doing and accept a higher price and more variety of produce.

Then imagine the second farm – on the same amount and type of land, but given over to just a three-crop rotation managed by a contract manager. The size of fields has been greatly enlarged to accommodate large machinery, which will sow, irrigate, spread pesticides and fertilisers, and harvest the crop quickly and efficiently. The varieties are high yielding, no cover crops or livestock are used and the farm sells just three products to distant markets.

Which of these farms would have the largest amount and diversity of invertebrates in the fields, soil and surrounding features? Given the huge variety of habitats, feed sources and wildlife corridors available, and the reduced chemical exposure, it is undoubtedly the first.

And which farm makes the best returns? Possibly the latter as they will have minimised the costs of labour, used cheaper energy and machinery, and sold their crops in a futures market to have an assured price. But it is a risky business. If one crop fails, they lose a third of their income. All too often, farmers here and overseas are squeezed by the big buyers with low prices and unfair demands, such as on delivery times and high-product specifications.

It is a stark comparison. Unfair you may say. But the contrast illustrates the choices farmers must make, often in difficult circumstances, and not always with the best or impartial advice from agronomist advisors, who are too often paid by the industries supplying chemicals, seeds and other inputs. Farmers having to use single-seed varieties are obliged to use chemicals to protect the crops and receive a relatively low reward for their efforts. Or they use antibiotics to treat pigs and poultry kept in close confinement. But we know diseases and pest pressure spread faster in monocultures and close confinement than in other agricultural systems. To get the levels of products required by the buyer, they also use fewer rotations, even though rotations can be seriously helpful.

This is all a disaster for invertebrates. They have too little habitat, flower plant food sources, have to cope with too many chemicals, and have little chance to move about to breed and find food as green features like trees and hedges are just disappearing.

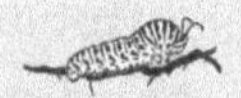

## Rebugging solutions for farming

Organic, permaculture and biodynamic farmers use different tools to manage pests to keep them at levels that are economically tolerable. They will keep fewer animals and pick specific livestock breeds and crops that are resistant to disease and pests, and they will encourage, often through providing habitats, biocontrols like wasps to eat the pests. So organic foods rarely contain pesticides residues in or on the produce and the impact of organic systems on invertebrates is far lower. Such farmers (using what's often called whole-farm agroecological systems) will also use complex crop and livestock rotations – the cornerstone of pest and weed management on many such farms and in many parts of the world.

You could say that in such farming the biodiversity is managed *within* the agricultural system as a form of *land sharing* (as discussed above), and contrasting with *land sparing* (where nature is protected by separating land for biodiversity outcomes from production). In a Dutch study comparing organic and conventional crop farming, earthworm abundance was two to four times higher than on conventional sites, though it varied between crops.[9] A major analysis of 94 scientific studies in 2014 showed that organically farmed areas have on average 30 per cent more species richness than non-organic.[10] Many farmers are showing it is possible, therefore, to produce food with far fewer or no pesticides.[11] It will ultimately just be different, more diverse foods that we need to eat to accommodate the changes. And it is encouraging to see that many 'conventional' farms are adopting many of these agroecological tools now, too.

*Integrated pest management*

Bees will use many tactics to protect hives, such as preventing mould by creating honey that has antifungal agents or using guard bees to check the hive entrance and alert the colony if it's under attack. When a hornet invades, Japanese honeybees swarm around their huge attacker creating a 'hot defensive bee ball formation' and vibrating their flight muscles. This movement creates enough heat for up to half an hour to kill the hornet because the hornet's exoskeleton is too strong to penetrate and sting it to death.[12]

Farmers, too, need to apply such a mix of tactics – called integrated pest management (IPM) – to reduce or eliminate the need for chemicals.[13] When done well, they may include more complex crop rotations, for instance, to break pest cycles, native or more resistant crop varieties bred conventionally for resistance as well as yields, good soil and nutrient management to keep crops healthy so they can resist harm, and use better forecasting and monitoring of pests and disease levels through good observation.

Using pest predators like the aphid-eating ladybird and lacewing is a great experiment in rebugging with benefits. In glasshouses, this is easy as you have more control but, at crop scale, relying on unstable natural processes is risky as farmers can lose the whole crop and therefore their income – especially precarious when they get such low returns from the market anyway. Plus they have other variables such as unpredictable weather to contend with.

Research is underway to assess the best and safest ways to encourage natural predators in the field. The summer I spent breeding leaf miner fly hosts was to do just that – assess ways to produce quantities of a host fly so we would be able to use them to breed a tiny wasp that is a parasitoid of the leaf miner fly in contained glasshouse environments. This was thirty years ago, so you'd expect more widespread use of these systems. Another study looked at the tomato – the second most consumed fruit worldwide (though eaten as a vegetable) – and its biggest pest in field production in Brazil, the pea leaf miner. They found that the presence of natural enemies, particularly a tiny wasp species, was among the most important factors for protecting yields.[14] As I explored earlier, such tools need to be used very carefully to avoid unintended consequences especially in field environments where encouraging predators (via providing habitats) rather than mass introductions would be the safest approach.

If it is so good then, why are farmers not using full IPM and biological controls, and turning to chemicals as a last resort? The main reason is often the policies and advice of chemical companies, farm advisors and government and a strong pressure to deliver high yields and uniform produce.

*Better policies*

We need more research and new policies designed to encourage a whole-systems approach via a combination of good independent advice, incentives and compliance requirements as well as encouraging consumers to buy more variety and not reject small blemishes or odd shapes.

We need to demand new policies to support farmers in using organic methods and IPM, and to ban the most harmful chemicals entirely. The European Union in 2020 announced a new 'Farm to Fork' and biodiversity strategy for the whole of Europe.[15] This focuses on chemical use and aims to reduce by 50 per cent both the overall use of chemical pesticides and also of more hazardous chemicals by 2030, with greater use of IPM and an objective for at least 25 per cent of agricultural land to be under organic farming management. An ambitious and pro-bug plan. I hope it gets done.

## The meat of the matter

Livestock play a big role, too, in the climate crisis, which also severely affects invertebrates. The methane-filled burps of sheep and cattle are full of climate-changing gasses, while emissions from livestock dung, forest clearance, heavy use of fertiliser on their feed crops and other land use changes all release greenhouse gasses. And globally we're eating more meat than ever.

I have been campaigning on the climate impacts of cheap meat since 2008, so I am relieved to see more awareness and research now. In 2019, the UN Special Report on Climate Change and Land confirmed the role of livestock in helping to drive catastrophic greenhouse gas emissions and said action was vital. The UN proposal included how shifting diets 'away from cattle and other types of meat' lowers the carbon footprint of a country.[16] The rapid global growth of a cheap meat system is finally being recognised for its huge negative impact on biodiversity, including invertebrates, as well as climate change and toxic nutrient pollution in rivers and coastal areas.

We can get livestock rearing right, but we need far less of it and to do it much better. The idyllic pasture we imagine for cows and sheep can be good if it is the right breeds, in fewer numbers and in well-managed and richly diverse pasture, ideally with trees, too. Dung beetles and many other invertebrates will thrive in such a field and I have visited several in England where the bugs are everywhere around you and the animals thrive.

Sadly, grass-based systems often are overgrazed and over-fertilised, which can cause significant loss of invertebrate life and so lead to loss of birds and other species. One study, for instance, on UK uplands assessed the impact of different densities of sheep and cattle on bug numbers and diversity.[17] Over several years, the trials showed that the total number of plant-dwelling invertebrates like spiders, bugs and beetles increased significantly with less grazing – and did better when cattle were included, not just sheep. Not only is pasture management clearly key, but the numbers and types of animals, too.

## The story of food loss and waste

The cockroach is a master of waste recycling, being a generalist feeder and sweeping up and using waste wherever it can find it. But there is a bug for almost every part of the food chain, from leaf litter (worms) to dead bodies (carrion beetles). We should be more bug-like and waste no food ourselves.

## Rebugging solutions for livestock

It may seem a long way from rebugging, but there is little of more importance to invertebrate life on this planet than a reduction in meat and dairy over-consumption and support for well-managed livestock systems. Meat eaters can take heart, however. If you need to eat some meat you can – few studies suggest we must all go vegan (though that is a useful approach). Eating far less and better meat and dairy can deliver the desired results.

In addition to well-managed pasture-based systems for cattle and sheep, we should feed chickens and pigs on waste products from the food system, (such as unused whey from milk production, used pulp from beer production, and unwanted fruit, vegetables and bakery products) and the unused parts of the plant or grains from crops. Eating less fish and avoiding intensively farmed fish and threatened species will also reduce the pressure on wild fish stocks that is hugely harming marine invertebrates and ecosystems.

Industry, government and investors should work together towards supporting a smaller, but better meat world, without investing in factory farming, and with proper support for extensive systems, combined with enforcing new dietary guidelines and standards for all public procurement of food (such as in schools).

## Rebugging solutions for food waste

From my experience, sadly, voluntary approaches, gentle nudges and corporate initiatives don't work: they have failed time and again. The very systems on which modern corporate food retailing is built – a fast, just-in-time and highly automated system based on centralised distribution centres, purchasing at low costs and marketing heavily so people buy more than they need – needs to be fundamentally redesigned to cut the waste and save the bugs.

We need strong rules with severe penalties for companies that are causing food waste in any part of their supply chain and that must include where they reject perfectly good farmer produce for spurious cosmetic reasons. We also need governments to provide citizens with information on how to cut out food waste and to compost what you can't use (with cheap wormeries or doorstep food waste pickups, for instance), and penalties for excessive waste.

I find the statistics on food waste staggering. Throwing away the crops or livestock produced is a massive, wasted opportunity to protect invertebrates. We could produce less and reduce the burden on the natural environment. We could share or spare more land. Food has become so cheap we don't think about wasting it.

In the UK, we threw away 9.5 million tonnes in 2018, mainly from the home and eating out. This is about £730 of food thrown away by the average UK family. Manufacturers and supermarkets have much less waste, but encourage wasteful habits through promotions, for instance, and cause farm-level waste by poor forecasting and rejecting produce for cosmetic reasons.[18] While accurate data is not always easy to find, the FAO estimated that globally at least a third of the food produced is lost (before it reaches the shops) or wasted each year. They are working to improve the data sources while also aiming to drive government and industry action.[19] This cannot come soon enough. Such an unsustainable and unethical waste of land, chemicals, animals, energy, water, effort, seeds, packaging, traffic and food.

## Local versus global

Finally, it is worth looking at *who* we buy our food from, and why. Buying direct from farmers who are working hard to protect nature and invertebrates is the best way to support their work. Farmers in the UK get an average of just 8 per cent of the money spent by consumers on food – the rest goes to processors, retailers and corporate shareholders.

But farmers need a bigger share in order to be able to farm carefully for invertebrates. And if we want to buy wildlife-friendly foods and follow bug-friendly diets, how can we send signals back to producers that we want changes and support them in providing that?

Farmers need a diverse mix of farmer-focused places to sell and buy from and I see this is growing in the form of farmers markets, direct sales online, box schemes, internet retail, co-ops, community cafés and community supported

## Rebugging solutions for food purchasing

We need to take back some control over where we can buy food from and that means supporting the kind of producers that help bugs. Finding producers more locally, and using organic and 100 per cent pasture-fed meat and dairy. Eating much more variety, mostly fresh and unprocessed, mostly plants, organic and fairly or directly traded where possible. Governments need to help this consumer shift through supportive policies and investment in regional, affordable food systems in order to reverse the global trends towards more processed foods.

Turning entrenched diets and the industry that feeds off it will not be easy. The Food Ethics Council in the UK, along with others, is working on a new way to encourage engagement in the food system beyond what consumers buy.[20] The simple approach of addressing people as 'citizens', rather than 'consumers', makes them more likely to care about one another, act collectively or actively participate in society. Currently, describing us all just as 'consumers' is deeply embedded in society and this suits those players keen to minimise societal interference in what they do. It drives policy the wrong way. If we can act instead as a 'citizen consumer', we

> can begin to influence what rules there are as well as the incentives to drive bug-friendly food production. Ensuring a right to food, including adequate incomes and welfare safety nets so people can access good food, is vital.

agriculture (CSA) that put consumers more directly in touch with sustainable farmers. Better Food Traders networks and direct sales to canteens and restaurants all create markets for the food produced in ways that help invertebrates. Those able to afford and have access to these traders or local and organic farmers are lucky – they can chat with producers and know that standards for better wildlife and habitat protection have been enforced.

But the market share of all these great initiatives is still too small and the big retailers and manufacturers dominate the food markets. Governments must control their activities better. But it must also invest in better trading systems to build a new vision for a biodiverse food supply and so farmers can build up more direct and shorter markets, and reach customers more easily. Shorter supply chains mean better communication and less chance of confusion or even contamination and loss of food quality.

## Your t-shirt has stains

I have written about the extraordinary design of invertebrates. For many bugs, from the huge rhinoceros beetle to the tiniest flea, one key element is a protective but flexible

## Bug body armour

One simple way many invertebrates protect themselves is in their body armour. Most adult insects, crustacea and other invertebrates have extremely hard exoskeletons, which protect them from damage and dehydration. Made of strong material called chitin (a biopolymer like cellulose embedded in a protein matrix), this sets as a hard, protective shell. This means it must be shed periodically as the animal grows. It is not as strong as bone (so suits smaller animals but could not support larger ones), but its flexibility comes in very handy for wing structure as well as external armour. Humans use chitin as fertiliser, as a food additive and emulsifying agent, and even for medical purposes such as artificial skin and biodegradable surgical thread.

The structure of an exoskeleton contains elaborate microstructures that are organised in several levels and it is both strong but flexible and adaptive to needs. This has inspired architects wanting to build tough structures that are also able to respond to external factors. Such properties can provide support in both compression and tension, increasing the earthquake-resistance of buildings in an active seismic area, for instance. In addition, through observing how the exoskeleton adjusts to the growth and

> the development of the insect, architects are able to mimic that capacity, and so using a 'building exoskeleton' structure on the outside of vulnerable buildings to retrofit, extend their life and adjust to new conditions.

outer skin – the exoskeleton. This outfit can be used as a weapon, as camouflage, as a means for transport, for communication and for attracting a mate. When bugs grow, they shed it and replace it with a new one. Some exoskeletons are so incredibly strong, for example the fantastically named diabolical ironclad beetle can take loads about 39,000 times its own body weight.

Most of us don't need that kind of armour, but we do have such an outer layer – our skin. It is impermeable, protects us and grows with us. But we cover it with clothes (obviously) and buy lots of other gear to do the job bugs have on hand – phones to communicate, cars to travel, guns to fight. We take much from the land to produce all these products, and this harms the bugs. We are buying more from the global marketplace than ever before with a 15 per cent increase in global per capita consumption of materials since 1980. We need to understand how these goods may be affecting the bugs.

## The impact of our clothing on bugs

Do we consider how the production of a cotton t-shirt or some polyester trousers may have affected the beetles where they were manufactured? The water used and the chemicals sprayed or used to dye the clothes – did it involve insect forest habitats being cut down or polluting the caddisfly streams? Are places bugless because of the clothes we wear? The extraction and processing of other non-food consumables like timber, paper and metal goods will also have a huge impact on the local wildlife and ecology.

The issues are in some ways similar to food production: how much land is used or converted into cropped land where it was once forested or used for other less intensive crops; the land use change; the diversity of crops; unsustainable levels of water use and pollution of water systems; chemicals used throughout the production process; and greenhouse gas emissions.

Shamefully, waste is again a big issue now. In the UK alone, around 300,000 tonnes of textile waste are put in the rubbish stream each year and either incinerated or put in landfill. Less than 1 per cent of clothing textiles is recycled.[21] As with food, this is a huge waste of the land, chemicals, energy and effort used to produce the materials. The average person in the UK now buys over four items of clothing a month. What would it take to make your clothing part of your rebugging challenge?

## White gold – but not for bugs

But some issues are specific to textiles and fashion. We rarely think about where our t-shirt comes from, but the cotton it is made with – 'white gold' as it is sometimes called – is big business. Its production requires vast areas of arable land, around 35 million hectares, largely in warm regions.

Important bug habitats have been destroyed already. Over 80 per cent of the unique Tugai forest in the Amudarya Basin by the Turkmenistan's and Uzbekistan's border, has been levelled to make room for cotton cultivation.[22] This is a vital habitat for rare species including seventy-eight species of invertebrates, especially butterflies, beetles and dragonflies, and the bugs in turn feed many other species including snakes, frogs, and rare and migratory birds.

More than a quarter of the 40,000 species of Neotropical (i.e. of Central and South America, southern North America and the Caribbean) butterflies and moths, nearly a third of the more than 440 species of Neotropical termites and a quarter of the nearly 550 Neotropical social wasps are found in the Brazilian savannah, together with more than 800 bee species. But cotton is also expanding there, where it is grown in rotation with soy and corn, to the extent that Brazil is expected to become the second largest producer (from sixth) and the world's largest exporter of cotton. As deforestation runs at the rate of 6,000 square kilometres per year and it is estimated to have lost approximately 50 per cent of its natural coverage already, many bugs will have already been lost to cotton.[23]

We know bugs can provide important pest control services, but in cotton crops these natural agents have been wiped out by the destruction of their habitats and through exceptionally heavy use of insecticides. Much of this insecticide is used to tackle the cotton bollworm moth, a hugely important crop pest, whose larvae feed on a wide range of irrigated crops in warmer regions. The caterpillars chew holes into the bottom of the cotton bolls and hollow them out.

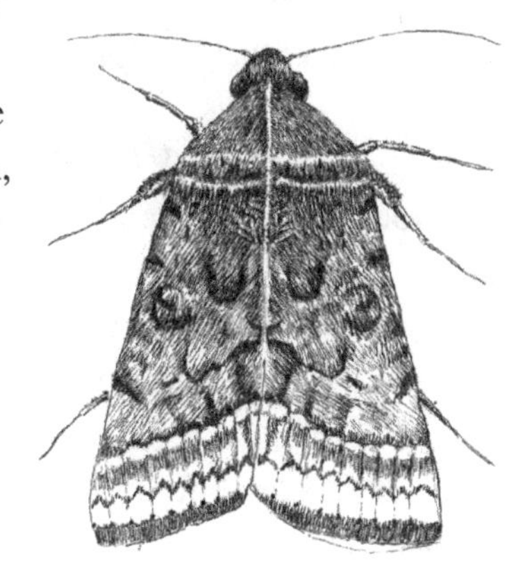

There is a huge army of natural predators of this pest moth including green and brown lacewings, assassin bugs, damsel bugs, spined soldier bugs, rove beetles, ground beetles, tachinid flies, phytoseiid mites, spiders and several dozen parasitic wasp species. But they all can be destroyed by insecticides. This leaves the bollworm as a major pest and growers have developed one strategy after another, with varying levels of toxicity and effect, over the past century to tackle it. A pesticide treadmill. It is one of the key reasons for the development of genetically modified cotton.

*A better way*

Turkey resisted the drive for GM cotton, using integrated pest management instead. As a result, it has lowered its costs with fewer inputs needed and increased yields since the 1980s, despite far lower pesticide use than other countries. Mali introduced alternative control methods and reduced its chemical use by 92 per cent. One Australian study shows a good example of successful cotton production using natural processes – pest control through natural pest management and reduced pesticide spray use, increased water-use efficiency and new fertiliser approaches. Cotton businesses have thrived while reducing their environmental impacts. So alternative, economically viable approaches exist.[24]

## The microplastic disaster

It's not just cotton textiles though. We also need to think about the bug impacts of other textiles like leather, wool and silk, which can all harm invertebrates in their production and processing. These natural fibres have advantages as they can be co-products of the food industry so the land impact may be smaller, but we should use them wisely. Leather, as a co-product of the cattle meat industry, can be associated with considerable harm. Wool is a crucial industry across the globe for sheep and goat farmers, is a great textile to wear and can be produced organically. But it can also harm bugs through over-grazing and the chemicals used to control livestock diseases.

But one type of textile now dominates the garment industry and is a major concern for invertebrates – man-made fibres such as nylon and polyester, made from plastics derived from fossil fuels. In 2016, sixty-five million tonnes of plastic was produced for textile fibres.[25] These textiles have one big advantage – they are not susceptible to larval insect infestation or other natural decaying processes. They are not a tasty food source for insects or fungus, unlike cotton or wool. This is also a big problem.

Microplastics are the tiny fragments of plastic released from artificial fibres and other plastic goods, and they are now found everywhere from the Arctic to European rivers and the deepest oceans. These fragments can harm the reproductive capacity, growth and survival of marine animals, affecting, for example, the invaluable work of invertebrates, like the caddisfly. As an adult, the caddisfly is moth-like but its larval stage is aquatic and has a telltale portable 'case', made of tiny stones or other materials, to protect themselves in fast-flowing water. Caddisflies play a key role in breaking down leaf litter in rivers, but when they ingest plastics, less leaf litter is decomposed.[26] Such impacts could be severe for all marine life.

It's inevitable that the smaller plastic particles, being of similar size, can be mistaken for food morsels by the critically important marine bugs called zooplankton. These tiny creatures underpin the marine food chain and play such an important role in regulating the climate through controlling numbers of microscopic plants – the phytoplankton. But if they are ingesting inedible plastic instead... And, scarily, we may be severely underestimating the levels of microplastic pollution in the oceans, with research suggesting that there are at least double the number of particles than previously thought.[27]

It is not just the seas but soils are also being filled with plastic. As worms move about and into new areas, they will refill soils with useful mix of microbes like bacteria, fungal spores and protozoa (single-celled organisms). They disperse them by their feeding and excretion habits or just through physical attachment to mucus or their body parts. As they do this, they will ensure that the new areas they move into contain a mix of soil microbes or fungi, which will then encourage plant growth by making nutrients freely available as they break down organic matter like old leaves and dead insects. The physical and chemical nature of the soil is also improved by the worm through their burrowing, casts and nests, which also create structures that can be habitats for smaller inhabitants. But we are now finding that microplastic pollution stunts worm growth and the health of plants and soil structures.[28]

Lugworms are mainly seen by us only through the curly mud casts they create on the sand surface of a beach. This worm lives in U-tubes it creates in the sand, filtering food particles and creating habitats for other species. Like its land-based relative, the earthworm, it is an important part of the food chain – a food source for wading birds – and you find the lugworm in most parts of the globe. When exposed to the kind of plastic microfibres now filling seas and other habitats, the worms' energy reserves were halved and the material spent longer than food in the gut, causing inflammation.

And there is more. As microplastics have a large surface area, they can also hold chemicals such as dyes, toxic chemicals and even diseases, and transfer these harmful substances when eaten. One study of the risk of disease in 124,000 corals from 159 reefs in the Asia-Pacific region suggested an increase in the likelihood of disease when corals were in contact with plastics ranging from 4 per cent to 89 per cent.[29]

Given the volume of microfibres now reaching the environment, it's clear we need to stop this pollution urgently.

## Rebugging our shopping

We buy so many things we don't realise can harm invertebrates where they are sourced – for example, products made of wood, such as furniture and paper, and palm oil, which cause immense deforestation pressure. And the mining of minerals such as aluminium, gold and lithium cause huge toxic waste, which will harm invertebrate land and water habitats.

Most of the stuff we eat, wear, build with, use and throw away matters to invertebrates, so we need to be wiser about the products we buy, where we buy them from and how we use them. Companies that take their responsibilities seriously will be buying from better and organic cotton farmers, for instance, or using recycled materials. There are new sustainable sources of materials made using fungal mycelium which are becoming available – such as fungal leather and fungal fabric – which will be a valuable alternative to harmful textiles.

Governments should ensure products are labelled accurately, that the most harmful products are banned and that reuse schemes are available. If a product is beyond repair, it should be recycled and the minerals and other components extracted and reused. We need political changes so that those choices become more acceptable, affordable and then the norm.

These changes will not happen without public pressure, sadly. We should be demanding that government put an end to harmful mineral or timber extraction, making industries report on their supply chains and banning products from illegal or harmful activities. In doing so they could be promoting the reuse of products and recycling where they can be reused. All too often strategies and policies fail to make the links between raw material production and distant consumption. This limits our ability to act and to find effective solutions.

Timber products can be a better and lower carbon-building material than fossil-based PVC or cement, but should only come from certified sources where they have been grown in sustainable ways or only taken from forests in small amounts, so preventing major forest disturbances. As with meat, we should be using less timber and from better and certified sources or recycled timber.

## Rebugging actions

Some easy actions:

- Cook from fresh as often as possible to reduce the use of over-processed foods and packaging – it has the added value of being healthier, too.
- Choose non-uniform, oddly shaped, sized and coloured fruit and vegetables! They taste just as good as the cosmetically perfect produce. Many pesticides are used solely to make the produce look 'perfect' and uniform. Tell your supermarket that they don't need to use pesticides to make it look perfect in the feedback forms they often have in store or online.
- Make every effort not to waste food. If you cannot eat it, try to compost it.
- When eating out, try not to waste food – ask for a bag to take leftovers home, and encourage restaurants and caterers to donate unused good food to charities and compost inedible leftovers.
- You could say the most sustainable clothes item you have is the one you already own, and the more you can reuse or repair what you already have the better.
- Buy sustainable clothes made from organic cotton and co-products of food production like organic wool from sheep or leather from cattle and make them last.

- Avoid buying plastic-based textiles like polyester, nylon, acrylic and polyamide but, if you do, wash them at lower temperatures with a full wash or in a fabric bag and at a lower spin to help reduce fibre run-off into water supplies. And don't use a tumble dryer if you can avoid it – it's bad for climate and plastic pollution!
- Timber, if produced sustainably and fairly, will technically be a carbon store – locking in the carbon it has absorbed as carbon dioxide so it does not contribute to greenhouse gasses in the environment – if kept in use forever, so buy from local suppliers or refurbished when you can.

If you can go a bit further:

- Where possible, choose seasonal, local food and consider switching to organic for some of the items you use the most. There are many online guides to what produce is available in what season.
- Buy organic and products certified from sustainable sources like Forest Stewardship Council (FSC) or, even better, recycled natural textiles, clothes, paper and other materials (linen, timber, furniture) because certification is not the main solution to end deforestation but does help us avoid illegal timber sources.

- Choose grass-fed meat and dairy. Well-managed, species-rich pastures, if not artificially fertilised, are often great for wildlife, will help retain water and soil, and can store carbon thereby reducing greenhouse gas emissions. Plus, 100 per cent grass-fed meat tends to use land that can't easily be used to grow other food.
- Only eat certified sustainable fish. More than 70 per cent of the world's fish stocks are overfished, depleted or collapsed. Making better seafood choices can help to ensure ocean health and sustainable harvesting – protecting the invertebrates that live there.

Big actions with big impacts:

- Think Before You Buy. Nearly everything we buy affects the environment, including the materials and energy it takes to make and operate each purchase. Ditch 'single use' and look for more sustainably sourced products and energy-efficient devices.
- Think Before You Throw Away. Reuse timber and paper. Reuse, repair or donate gently used items whenever possible. If you need to throw something away, check with your town council for the best way to recycle or dispose of unwanted items.

- Try a new diet. Eat organic food or food where you are sure pesticides have not been used (for instance because you grew it!), eat less and better meat like organic and pasture fed, eat less junk food – a waste of crops – and cut the plastic to avoid microfibre pollution.
- Cut out damaging products, like palm oil and soya, and genetically modified foods as these are mostly designed to work with pesticides. You may find palm and soya in non-food products like soap and shampoo, too.

— CHAPTER 7 —

# Putting bugs into politics and the economy

That this book about invertebrates and rewilding also explores power, inequality and consumption may seem strange, but I hope you will keep reading. I could not write a book like this without talking about these issues. They are all part of the system, which needs fixing. And it cannot continue. If we really are to rebug, we must change how decisions are made. If we do not tackle the deep-seated, often hidden, political and socio-economic drivers of harm, then we will fail. For invertebrates to thrive they need system change.

We need to act as citizens and political actors, not just as gardeners or farmers or consumers, to help protect invertebrates and the habitat they need to thrive. For too long, campaigns have focused either on protecting special sites or specific species. They are important but have often failed to get at the heart of what is causing the invertebrate decline. These drivers are often hidden, and the players powerful and extremely smart at being able to continue to do what they want.

Bugs in complex, super-societies, such as bees and termites, have clear roles and responsibilities and fantastically

### Termite cooperation

They may have a reputation for damaging homes and crops, but termites are super-cool, orderly insects that define the very image of highly successful collaborative working. The colony members – of different ages – all know what their role is in the complex system within the nest, which is also highly complex in construction. They have intricate chemical and vibration signals to organise a robust defence of their nest and, when needed, members will sacrifice themselves for the colony. In a French Guiana species, ageing workers literally explode, showering an enemy with toxic blue liquid they have stored on their backs – true self-sacrifice. Other workers won't breed but instead care for the termite young, collect food or build and guard the nest. Through altruism, these colony members help to increase the chances that the genes they share survives. An impressive example of working together for the common good.

effective channels of communication. They each know what they must do and leadership is managed to ensure the colony is well guarded against invaders, that the workers, babies and queen get fed with the right nutrients and that the nest structures are maintained.

Bugs will also work with other species for mutual benefits. The wood ant encourages helpful worms into their nest and lemon ants kill off unwelcome plants that could compete with their chosen tree. Given that these are sometimes many millions of individuals, this seems a perfect example of fantastic governance using complex communications.

Yet we humans have developed systems of governance that appear to be taking us fast in the wrong direction and we have too little control. And invertebrates are some of the best indicators that our relationship with, and stewardship of, the natural world is out of kilter. Their diversity, adaptability and sheer numbers mean they, of all creatures, should be able to survive, but even they cannot withstand the multiple stressors.

So, if the evidence is there, alongside the thousands of organisations at global, national and community level, working hard to put things right, why are we not getting it right? Why are we still making decisions that cause such harm? Why is the evidence growing of a major crisis in invertebrate populations?

The difficult truth is that vested interests hold far too much power on decisions, meaning the right change or the systemic changes are rarely made. Those vested interests even include huge, largely invisible asset management companies. Who knew? To compound the problem, when solutions are proposed, they are too often focused on one part of the system or a single challenge, such a source of pollution or damage, when the root causes actually lie elsewhere – in policy drivers or business practices. Issues of power and influence are too often hidden. We also tend to avoid addressing entrenched systems in society, such as poverty and inequality, or attempt to properly tackle over-consumption. Because these are all hard to discuss.

They may also seem far removed from rebugging the planet. But if we do not get the decision-making right, any

## Mutual aid

Lemon ants live with a tree known as *Duroia hirsuta*. The plant provides them with a nest and in return the ants protect the plant from herbivores and clean the surroundings, preventing other plant species that might offer competition for nutrients, light and space to grow nearby. The ants are named after their flavour – if eaten, lemon-flavoured citronella pheromones are released to warn nearby individuals. The ants bite any other plants trying to grow and inject formic acid, effectively a weedkiller, into the young plant, killing it. Large clearings in the rainforest, called 'devil's gardens', show the presence of this relationship as few other plant species can live there.[1]

Often such mutualism is called 'obligate' – the species co-evolved and will not survive well without the relationship. We know, not only from the science but from our own eyes, how broken the relationship is between humans and the natural world – from widespread deforestation to species extinction through hunting. We have taken far too much out without understanding these complex relationships and our very survival is threatened as a result.

actions we may take as individuals and communities will not be enough to change the status quo. Big food and agro-chemical businesses, and large retail and finance industries, hold sway over so many of the decisions that affect invertebrates.

Like the ant colony, it is a complex web of interdependencies and interactions we need to address when it comes to power and governance. It may seem like an obvious point, but we need whole-system approaches led in the interests of society as a whole and the interests of nature, sharing the space not dominating it.

And currently that is generally not how politics work. Or economics.

I suggest that there are three areas that are critical to understanding how we are, where we are and why that matters for invertebrates: poor governance and politics; inequality and poverty; and runaway consumerism. Truly rebugging the planet will depend on tackling these three, however difficult they seem.

## Poor governance and politics

The easiest way to measure for impact, both positive and negative, on invertebrates is assessing land use and land use change. So, looking at those who make decisions on land and who oversee its management is vital to understand the biggest influences on invertebrate survival and well-being.

Are our governments representing the collective view and best interests of the colony, so to speak, able to adequately protect the animals and plants? Or is it the major corporations and investment landowners who wield the most power over decisions? All too often, the latter holds the most power and, while some corporations and investors are interested in protecting biodiversity and the environment, that is not their primary purpose.

Farmers, foresters and other land managers, who may want to make the best decisions to protect ecosystems, often cannot because the corporations they sell to do not pay enough or require specifications that make that impossible. Remember the tale of two farms? The deal the farmers get is a big part of the problem.

A few facts illustrate who is really in control of the land, looking at global food traders first.

- Agricultural commodity traders are now the most powerful companies in the industrial food chain. Six companies control most of the global food trade and earned $376.9 billion in 2018. Revenues of the world's largest, the privately owned Cargill, were $115 billion.[2]
- The impact of the countries and corporations managing this trade on market stability and prices, and therefore on production itself, is huge. Decisions and profits are concentrated in the hands of just a few corporations, while farmers and farmworkers take on all the risks of production with little protection from abuse. Many have had to get big and intensive or get out. Bad news for the bugs.
- The fact that soybeans, wheat, rice and maize account for three quarters of the calories that people consume globally means a too high level of specialisation and homogenisation – all bad for invertebrates, as I explored earlier.
- The processed or branded food we buy from shops – as opposed to buying fresh – is produced by some of the largest food companies, who have grown ever bigger through mergers, swallowing up smaller national companies. You have an illusion of choice when you

are in fact buying from a small handful of corporations. And when did we really decide we needed thirty varieties of biscuits?

- Just ten companies (Nestlé, PepsiCo, Coca-Cola, Unilever, Danone, General Mills, Kellogg's, Mars, Associated British Foods and Mondelez) control almost every large food and beverage brand in the world. They make some of the largest profit margins in food chain.[3]
- Just thirty global supermarket chains control a third of the global retail food market. They demand uniformity of product – be it milk, flour, palm oil, beef or tomato paste – they pay low prices for these raw materials and buy huge volumes of it. More pressure on the farmed land and so on the bugs.

Unlike a well-running complex ecosystem, this is totally out of balance. The huge concentration of buyer and seller power squeezes producers at one end and consumers at the other. It is why I have campaigned for many years for better rules to control the market share and buying practices of the big companies. We are still sadly a long way from taking back control to protect ourselves and nature.

Using the analogies of complex mutualism or the 'super-colonies' of social insects like ants or termites, we could look at the remarkable ways resources, including heat, food and water, are shared, used wisely and provide food for all. The human tools of production – the chemicals, the seeds and other farm inputs – are often not in the hands of those who need them.

For instance, genetic resources – the seed varieties and livestock breeds – underpin most food, cotton and wider agriculture, forestry and fisheries production. Many small-scale farmers have always saved and shared their own seeds

## A bug-built skyscraper of (relative) harmony

Coral reefs are extraordinary neighbourhoods of invertebrates and other animals living, eating, hiding, hunting, breeding and dying in one vast structure. This is formed by a bug – tiny individual coral polyps (in the same group as jellyfish) using the minerals in seawater to make limestone exoskeletons. These exoskeletons fuse to form huge structures and a home for several thousand species of animals and plants. The polyps themselves are colourless, but take on the shades of the algae they host inside them, which use the polyps for shelter and their waste products for photosynthesis. The algae, in turn, produce oxygen and carbohydrates, which is used by the coral to grow and build. Worms, shrimp, spiny lobsters and crabs are just a few of the invertebrates that call this skyscraper their home, and sea urchins often repay their hosts by vacuuming up the algae that may otherwise smother the reef. A dynamic, self-governing, mutualistic society in action, with a tiny keystone bug at its heart.

from year to year and breed in traits such as disease resistance or taste qualities over many years. But the farm sector that sells into the major commodity markets that produce the bulk of food in shops often have no such choice. They are obliged, through contracts and specifications, to use specific genetic varieties of seed or livestock, which they have to buy in each year. Concentration in farm genetics is happening fast: in 2018, over 60 per cent of global proprietary seed sales was controlled by only four firms, and just three companies controlled virtually all the world's poultry breeding stock.[4]

The growth in use of genetically modified seeds should be a major concern to us all because of the impact on pesticide use, fertilisers and monocultures. One recent application by Bayer was for a corn variety engineered to be resistant to five pesticides at the same time.[5] A serious chemical cocktail for the invertebrates above and below ground. A bug species that did this to its home would soon be wiped out.

Talking of pesticides and fertilisers – these are also tightly controlled by a handful of companies. In 2017, just 4 companies controlled 70 per cent of agro-chemical sales worldwide and 10 companies controlled over 50 per cent of worldwide sales of artificial fertilisers.[6]

Finally, why did I mention big investors earlier? Many of the huge companies involved in agriculture trade and inputs are now actually owned by major asset management companies that invest their clients' funds wherever they see likely growth. One assessment calculated that at the end of 2016, the world's five biggest institutional investors – companies who have nothing to do with land or food – owned shares in the world's five largest seed companies.[7] These companies are so far removed from the business of soil, it is impossible to imagine they would act, or be compelled to act, in any way to protect bugs. To take the ant supercolony analogy further, it's as if some beetles decided to take all the ants' food supplies even though they cannot eat or use them. Money accumulation is hard to eat.

But their ability to affect farm, land and forestry decisions, to determine the incomes of land managers and influence what is produced has grown so significantly over the past century, it is inevitable they now affect invertebrate diversity globally.

*To rebug properly, we need to take back control*

And to do that we'll need to counter the huge reserves of cash these powerful companies hoard to finance their lobbying against regulations that curb use of their products, or strengthen environmental or worker protections. In 2019, in the US about $139 billion was spent on lobbying government by agribusiness, using at least 1,149 lobbyists to reach the top levels of political parties to ensure any changes to legislation and regulation favour their interests.[8]

These lobbyists' influence on decisions in areas like chemical authorisations and controls is immense. Many pesticides banned elsewhere on public health or environmental toxicity grounds are still authorised for use in the US.There is a strong environment movement in the US working hard to protect nature, but they have nothing like the same resources. Europe has its fair share of corporate lobbyists, too. They infiltrate influential advisory groups, holding meetings with top officials, funding parliamentary industry groups and hiring expert lobby consultancies. But who lobbies for the bugs' interests?

There is a sorry tale to tell about how the agro-chemical sector has worked, and spent millions, to influence decisions on pesticide rules.[9] One of the biggest battles in the corridors of European power has been around Monsanto's best-selling herbicide Roundup, which contains glyphosate – a chemical compound suspected to cause cancer in humans and yet is the largest weedkiller used globally. It is used on crops that have been genetically engineered to be resistant to glyphosate, so extensive spraying is common. It therefore has a huge impact on invertebrate food, habitats and environment.

The story of glyphosate shows how well companies like Monsanto use all the tricks of the lobbying trade, while working to stay almost entirely below the radar. Their campaign was run by a group called the Glyphosate Task Force, usefully located in the European Union quarter of Brussels. When the authority responsible for renewing the licence (which means it can continue to be sold) produced its assessment report, it was found that dozens of pages of their paper were identical to passages in an application submitted by Monsanto on behalf of the Glyphosate Task Force. Similarly, partial bans of those bee-harming neonicotinoid insecticides, introduced in the 1980s in Europe, are constantly opposed by the makers and, elsewhere, neonics are still widely sold, causing ongoing harm to invertebrates across the globe.

We all need to be involved in governance at local, national or international level. We need to act as worker ants together protecting the colony against harm. It is probably the biggest challenge we face – to ensure that decisions which governments or those in power make protect the interests of nature and wider society. They clearly need to work independently of big business, so we, as citizens, need to be involved in that process to check for signs of weakness, to be ever alert like the soldier ants. That also means the public supporting and working with organisations that are investigating the issues and holding politicians and institutions to account.

## The impact of inequality and poverty

It may seem unconnected, but inequality drives harm to invertebrates in several ways. Land is again at the heart of the story – inequality of land and its use. Land ownership is highly political and often embedded in historical issues of ownership and colonisation. We should not shy away from these subjects though – they are key to driving better outcomes for invertebrates, however politically tricky.

Access to land, or influence over who manages that land, will be a key part of both rewilding and creating land uses that promote invertebrate health. Mismanagement of land, such as mining, deforestation, depletion of soils or toxic contamination, ultimately affects poorer people, including smaller farming communities, most as they are the most vulnerable. They lack the legal resources, ownership or tenancy rights or political power to stop encroachment on land or land grabbing (where their land is opportunistically bought or leased, often by domestic or global companies, or even governments). The lack of power also prevents them from being able to stop contamination and damage to their land, water, nature and soil resource. Examples the world over show that these situations may lead not only to environmental harm but also displacement of whole communities, landlessness, hunger and greater poverty.

By contrast, when someone's livelihood and ability to feed and school their family depends on being able to grow or harvest produce or rear livestock on land, they will rely on the natural resources surrounding them. They will care for it. One great example is the use of the tree-living weaver ants by tree croppers in Asia. The ants are such good predators of the pests of tree crops that weaver ant colonies, made of hundreds of silk-sown nests hung in trees, are nurtured by forest farmers in many regions.

Years of growing and tending the land and forests will have given them a clear understanding of how the natural systems function – the trees, the water resource, soil nutrients and pest predators – and how to protect them. If the land is taken away, or a new regime imposed on the community, this can harm the delicate ecosystems that have been built up over generations of careful cultivation. Over the last century, the removal of such communities has accelerated massively, including pastoral or forest communities driven out to clear timber or to create pastures or cropping or resource extraction, often for export.

### *Land is life*

I saw the impact of land grabs creating inequality when I visited small-scale farm forest communities in the Chaco region of Paraguay.[10] I could see the vast fields of Brazilian-owned, genetically modified soya plantations right next door to their own forest-based smallholdings. For centuries, communities there have grown a mixture of crops and animals within forested areas to feed themselves, their family and community. In the past decade, some had been bribed to give up their land to soya barons and these lands were then bulldozed and left devoid of insect or much other life but soya plants. The soya bean and oil were destined for export to be fed to industrial pig, poultry and dairy units in Europe and the East.

The farmer explained to me that 'land is life' and how he wanted us to spread the word about how their diverse, ecologically sound system was being threatened by our cheap meat diets. He showed me his tomato seedlings, the third ones he had planted that year in that spot because the herbicide sprayed on the soya crops next door had drifted onto and withered his previous two crops. He was stoic in the face of such destruction and I felt wretched.

A second cause of invertebrate harm resulting from poverty and inequality concerns the drive to lower the cost of production. The downwards pressure of poverty on income and wages leads to a demand for even cheaper food, which inevitably leads to a drive to lower the direct costs of production – to get ever higher yields from the land and livestock. This means the kind of intensive, chemical-based farming systems I have described earlier, which are most harmful to invertebrates. Inequality and poverty can also be created in turn by mechanisation, which reduces the number of workers needed on farms driving rural workers into urban areas where, without adequate policies, they too end up in poverty. It is seen as progress – but at too high a cost.

## A more equal relationship with nature

The double whammy is that the *social* inequalities are also now causing *natural* ones. I have little doubt now that the environment, economics and politics are intimately interconnected, through human interactions with the natural environment and with each other. Given that almost half of the world – over three billion people – live on less than $2.50 per day and three out of four of these live in rural areas: poverty, inequality and environmental protection go hand in hand. We don't share well.

I could also stretch this word to cover inequality between us and the natural world. Our relationship with nature is unequal – we assume we should have total dominion over the systems and the creatures with whom we inhabit this world. Unlike with those forest communities working with ants for centuries to protect their crops, our power to 'control nature' is not balanced with the knowledge, skills and desire to do it well and for mutual benefit.

Take our desire to eliminate any bugs from our gardens. Why would we want to do that, or make lawns so uniform and spray insecticides on flowers? We create huge cereal and other crops using chemicals and genetics to remove the bugs just so we can feed livestock in covered sheds. We don't need that much protein, but we take the land from the bugs anyway. That is an unequal and unnecessary set up, but it suits those trading cheap goods across the globe.

Unlike insects, which when left without human interference live in balance with plants and other species, we can mostly and accurately be described as a deadly parasite. This has created a situation of harm to both nature and us and is something we desperately need to reverse.

With human inequality, we also now know that there is a deadly link between inequality and biodiversity loss.[11] Researchers have found that the number of species that are declining or threatened grew substantially in relation to an indicator of inequality called the 'Gini coefficient'. The Gini coefficient is a widely used statistical tool to show the distribution of income within a nation or group of people where zero means perfect equality and 1 shows perfectly inequality. Researchers have found that as inequality grows so does harm to biodiversity. There are complex issues at play, as environmental harm also increases inequality as discussed above, and more research is needed. But the links are clear. Inequality drivers may include behavioural aspirations that lead to overconsumption and, as the rich get even richer, more resources are needed to supply the rest of us with what we need, so leading to further encroachment on nature. There is habitat loss related to low-income communities forced to cut down forests for land or fuel. Additionally, inequality may hamper the collective action needed to protect environments, and deep inequalities detract from environmental policies because of pressure to deal with poverty first. Richer nations and the richest in society have the power to extract the most from nature – often irresponsibly.

So after controlling for other factors such as population size and biological and physical conditions, researchers have consistently found that the relationship between inequality and loss of nature persisted.[12]

This is strong stuff. To help the bugs we clearly need to help ourselves, too, and this is getting through to powerful institutions now. In 2020, the UN announced a commitment for protecting biodiversity and part of that commitment was about putting human rights into the world's biodiversity agenda. They were clear we will only achieve the crucial biodiversity goals, including for invertebrates, by major and deep changes in economic, social, political and technological systems. To 'bend the curve of biodiversity loss, we need to bend the curve of inequality.' Quite a striking statement and one now agreed by most nations across the globe.[13] If we followed through on this, the farmer in Paraguay would not have his land snatched because we'd stop using so much of it for cheap meat.

## Runaway consumerism

I have talked about how invertebrates rarely waste anything. What they need they find, no more, no less. But we have gone so far from a good model of consumption that matches sustainable production. Biodiversity loss has been driven by a spiralling increase in our consumption. The impact is multiple: more resources taken from the land; more pollution from production; more greenhouse gas emissions; more wasted products often further creating pollution; and the extinction of more species. We also consume too much for our health, with two billion people globally now overweight or obese. Scientists agree that we can only effectively tackle these huge problems if we change the lifestyles of the affluent and this is a problem for many market economies.[15]

To tread more lightly on the land, and save space, clean water, air and soil for bugs, we do need to act as individuals

## Domesticated bugs

Huge oak trees are home to thousands of invertebrate species, but one interesting association has been revealed only recently between brown ants and the giant oak aphids, which has been hiding in plain sight for centuries. The ants milk the fat aphids for sugary honeydew, but tend them like cows, moving them to the best spot on the tree for sap to suck and keeping them in shelters made of lichens and even bits of beetle. When disturbed, the ants will rapidly move their 'flock', carrying the smaller individuals in their jaws and urging the rest into safer shelters. A hidden symbiosis in action.[14]

– buying less and reducing waste, choosing sustainably produced goods where possible, and so on.

But this conflicts badly with the demands of a growth-driven economy, which urges us, with sophisticated, multi-million-dollar marketing programmes, to buy more and spend more. Politicians are still driven by an obsession with gross domestic product when that product is clearly driving harm to the natural resources we need and which creates a hugely unequal society, which further exacerbates bug harm. Tackling this needs political rebugging actions, more solidarity with each other and pushing for a different economic paradigm, one that is not based on unsustainable growth.

## A better future

Primarily we need to ensure people have access to, or the means to acquire, the natural resources that they need to make a living and which are essential to their livelihoods. That means stopping extractive agendas and land grabs by global corporations. It means development agendas that are based on solidarity, a fair transition and led by affected communities. They need to be determining the kind of environmental management needed and provided with the economic incentives for doing so, if the market or other conditions do not.

We should be providing compensation to communities for conserving or managing crucial resources such as forests, and providing employment in that work. Investment is needed in those natural resources and agricultural technologies with environmental benefits, and in promoting low-risk, low-debt, resilient production in poor and marginal areas.

One great example where this kind of approach is starting to work is in cotton projects in several African countries. As previously discussed, cotton is hugely difficult to grow but in high demand. Pesticide Action Network has worked in Benin and Ethiopia for more than twenty years, supporting thousands of farmers to grow organic cotton. They work with local groups to demonstrate the benefits of good crop husbandry, integrated pest management techniques and soil improvement without the need for expensive and hazardous pesticides, using five core principles: direct training and support (including farmer field schools); establishing village cooperatives; providing alternative pest management techniques; empowering women and girls; and improving food security. Their project results have been impressive, with communities getting far better returns for their organic cotton and the natural ecosystems, including the invertebrates, nurtured. A win-win scenario.

## Rebugging solutions for taking back control

Tackling the issues outlined in this chapter is not easy and it may feel some way from rebugging. But it is not. It matters for the bees and ants and the springtails and rotifers. There are many organisations and movements who have been working for years to take back control, and to reduce poverty, inequality and unsustainable consumption for the benefit of society and nature. But they need more of us to use democratic structures, mobilise the legal system and build support for a new economics and better rules to make the changes needed.

So I suggest there are four key areas needing action: democracy, new economics, corporate accountability and international cooperation.

1. As never before, we need fully democratic, transparent and inclusive societies that we can all be active in. Progress has been made and some systemic changes delivered via democratic processes. In the UK, the hugely important Countryside and Rights of Way Act and the Climate Change Act both came about because of individuals lobbying their parliamentary representatives alongside legal and lobbying activities of organisations.

Achieving action to protect communities from toxic waste dumps in the US has been achieved through legal means. These systems are always under threat as politicians seek to undermine them in favour of their own ideological interests or to benefit their buddies and lobbyists. But progress has been made, however slow, and rebugging can only happen if we choose to act and demand action from politicians and governments.

2. We need to rethink economics. It currently operates as if life support systems are infinite – that we will always have pollinators around to work for us, even as we poison them and remove their habitats. The logic that puts fossil-fuel extraction above action to curb the greenhouse gas emissions that are literally destroying ecosystems we need to grow food. Unsustainable behaviour must not pay.

   This must be a just approach and ensure those on lowest incomes and consuming the least have a decent future. Happily, new economics thinking is gaining ground, starting to infiltrate the education systems where the old-style economics is taught. One useful framing is around 'doughnut economics', developed by a UK academic called Kate Raworth, which presents twelve types of social standards we should aim for (such as peace, education and health) alongside nine planetary boundaries which we need to keep under (such as biodiversity loss and phosphorous pollution), to avoid unacceptable harm and potential tipping points in nature. By bringing social and environmental concerns together in a 'safe and just operating space' it forces governments and economists to measure outcomes and design solutions that work for all. An idea whose time has most definitely come.[16]

3. We need far stronger corporate accountability – companies to be legally liable for any harm they do to bugs (or people) and to recompense them when harm is done. The clothing company Patagonia is an example of what can be achieved. It is committed to activism and lobbying for nature, funds indigenous communities working to protect lands and produces goods with far lower impacts on the environment. And it is honest about what it fails on. There are smaller companies, like Riverford in the UK – an organic food company that works hard to assess its environmental impacts and is employee owned – and local, smaller-scale companies who source organic foods locally to sell to people locally; for example, all the organic veggie box schemes, farmers markets, and UK shops

such as Unicorn in Manchester and HISBE in Brighton, as well as the Better Food Traders network.[17] These seem more bug-like in their community approach, the sharing of risks and rewards, and their low impact. We need copies in every borough in every town.

4. Finally, international cooperation is fundamental to getting much of this right and securing a new vision that will protect invertebrates. Bugs travel, air pollution drifts, seawater circulates. Many national boundaries are irrelevant when it comes to natural systems, as we've seen from the spread of the Covid-19 pandemic in 2020. We need to strengthen the power of global treaties and institutions to tackle these huge problems and make sure they are immune to undue influence by corporations. We now have the awkwardly named Intergovernmental Science-Policy Platform on Biodiversity and Ecosystem Services (IPBES) to assesses the state of biodiversity and of the ecosystem services it provides to society. This must succeed in convincing decision-makers on a course of action to reverse bug declines.

Many people will find taking action to address these drivers hard. How can we, as individuals, control the big agrochemical companies? What is our role in tackling inequality and how can we influence decision makers? I have some tips for you below. Unless we do some of this, we will not make the changes needed to reverse the huge declines in invertebrates.

- One of the most important things you can do may be how you vote in elections – at local and national level, and how you engage with elected representatives. Getting active politically should not just be for NGOs and committed activists. Anyone can be a spokesperson, educator or role model in the community or take part in a movement. And no one needs to do this alone (see chapter 9 for how to get help).
- Pick a key issue you feel strongly about – such as a company you feel should be doing things differently locally or globally, or the protection of a local tree, or stopping the lobbying power of pesticide companies – and focus on that. Join a campaigning group and become an active local campaigner, drawing in others to help you. Start a new group if you need to.
- Join organisations (chapter 9 lists many) working to address inequality and in solidarity with communities across the globe, such as the food sovereignty movements, workers organisations like unions and those working

for human rights – the right to food, water or land. If you are interested in economics, see if you can persuade your council or city to adopt the doughnut economics ideas in their operations – some already are, like Amsterdam.[18]

- Reduce your footprint on the planet where you can by following the tips in this book on reusing, recycling, and buying less and better. And encourage others to be part of the movement to reduce consumption.
- Always be a real 'rebugger' and bug the politicians at local and national level. Often. The big changes we need take major government action and that will only happen with a strong movement for change (see 'How to start lobbying for change' on page 172).

— CHAPTER 8 —

# Imagine a rebugged planet

I am determined to conclude by painting a new vision, a bright alternative to my opening vision, of what the future could look like if we did rebug our lives and the planet we live on. What would it look, smell, sound and feel like?

Imagine if all lawns and parks just had strips cut for us to sit on or walk through, and the rest was left more often to grow like children's hair in the Covid-19 lockdown. All sorts of grasses and flowers would appear. Gardeners and park keepers could experiment with timings – when to mow to get particular grasses and plants flowering and then shedding seeds.

More and more species of birds and mammals would thrive in the new habitats that were full of dinner for them and their offspring. Hedgehogs, for instance, would benefit from new urban habitats and food (and no toxic slug pellets). I used to see hedgehogs all the time in my urban new build in the commuter belt and I would love to see that once again.

To me, one of the clearest outcomes would be the striking increase in the diversity of plants and features all around us. Some may grumble that it looks messy. But I think they'd soon come round to the joys of flowers on verges, of uncut grass, of more hedgerows, shrubs and trees, and less bare soil.

Gardeners liberated from the tyranny of the perfect lawn would have time to deal with the pests and plan those biological tricks to tackle them. The competition for the most perfect lawn would be transformed into that of the most diverse grass sward... I can see a big change in school grounds where green spaces, growing areas and more trees are used to educate children about the natural world.

Beauty would be everywhere. Little bits of greenery as everyone starts to rebug their homes, towns and villages, and workplaces. Imagine having grass full of flowers and buzzing with life wherever you walk to get your lunch or in the playground.

As we start to rewild where we live, we will see the mysterious ways of nature unfold. More creatures to greet us as we step outside and walk down the street. Not just bugs, but birds and other creatures would flow in, encouraged by more bug food and flowers and the new green spaces that are rich with diverse plants and living soil.

All our senses will be touched by this rebugged world. I can imagine the smells created by more flowering plants across all parts of the urban landscapes, the richer colours, and sounds of a more living urban and rural community around us.

Bugs are enticed to come back – from the smallest window ledge to the largest park. As you walk down streets edged with wildflowers, you spy lawns no longer a single green but every verdant shade and full of butterflies and hoverflies.

Farmed fields are a mosaic of diverse crops with native livestock and hosting many wild and messy bits in between, full of insects and spiders. And as soil becomes ever healthier, we see the coloured array of dung beetles and worms filling the farm soils, busy hoverflies and bees pollinating as they fly among the crops, and predatory wasps emerging from the hedgerows. Our eyes are caught by a shimmering picture of fly larvae making the unpolluted rivers swell with more fish and emerging mayflies caught by swooping swallows.

Rural and remote areas that have been rewilded are now extraordinary, alive with nature doing what it does best, routinely baffling conservationists with the vertebrate and invertebrate inhabitants and visitors, and bringing joy to many. Major rewilding projects will multiply because rebugging shows what is possible and there is more space as we waste less. Letting the invertebrates do what they do well allows the rewilding to happen. It is already taking place.

We would see more bugs on the windscreens.

---

It is a giddy-making picture. As a scientist and campaigner, I am not usually one to indulge in unrealistic dreams. But writing this book has unearthed expectations.

Let's go further. If we are genuinely more bug-like in our everyday lives, we could be wasting nothing, consuming only what we need, reusing, repairing and sharing stuff, eating more diverse foods, working as a social colony (a bit more like the termites) to live well in our space. Imagine if we truly changed what we do, buy and wear, eat and use, and how we collectively work together to protect this shared home. And if that started to be reflected in new government policies so big changes started to emerge.

In this vision, farmers and other land users would find decent livings through both the marketplace, as new rules ensure they are decently rewarded by retailers and restaurants, but also supported by the taxpayer for doing and growing and managing things differently, for bugs and for nature.

The next part of my great vision would be a change in what we eat and what we buy. Diets would be more diverse, based on more seasonal and local varieties of grains, oils, vegetables, pulses, fruits and meats. We would eat far less but far better meat, so getting the nutrition without the excessive land and water use and biodiversity loss that factory farmed meats cost.

Like the very efficient ants who waste little, nurture their gardens and work with other bugs and worms to provide nutrients and water, we would start to work together to waste far less food, recognising it for what it is – the most valuable purchase we make.

How cool would it be to reverse the fast-fashion trend as quickly as it came in. We should make the toxic throwaway culture as unpopular as, well, slavery. I do not use that word lightly. There are more slaves in our modern world than in previous centuries, including in the clothing and food industries. They are just hidden in dense supply chains that also hide the ecological destruction. Instead, we would reuse, rework and recycle so that far less land or fossil fuels would be needed to create textiles. More space for nature and the bugs!

Chemicals that kill bugs would still exist, I suspect, in my rebugged world. But they would be tightly controlled, their use monitored and enforced through limited licences. We'd still need to manage pests and diseases. But the more natural

balance there is, the more biocontrols we nurture and the more mixed environments we create, the more we will lessen the threats posed by damaging invertebrates.

We'd need to learn to share 'our' space better. Will we put up with moth holes and termites? Perhaps not, so we'd need to learn how to keep them out. Finding better ways to protect materials we need and to keep out bugs through physical barriers when it's necessary. In this new world, where we do not bestow on ourselves an inalienable right to deplete and destroy nature, we'd work, play and live differently because we'd have and need less stuff.

Climate change threatens the bugs and us alike and if we don't cut fossil fuel emissions and emissions from land use, we're all in trouble. In my rebugged future, we will have ended fossil-fuel use and invested instead in renewables and energy efficiency. We'd be more able to manage the changes ahead as bug-filled soils, along with more trees and bushes, allow us to continue to grow food. Carbon- and biodiversity-rich forests and wetlands will be undamaged and can continue to do what they do so well for nature and the bugs, the climate and us. Indigenous forest communities would be able to manage them as they have for millennia, ensuring the harvest only of what is surplus to the needs of the forest system. Eventually we will become like the huge invertebrate supercolonies, working better together, sharing the resources and taking up the space we need, but no more than that.

You may think this vision seems far-fetched. We can't go back to an era when bugs killed millions and destroyed crops and buildings. There are too many people and too much development to turn back on. And people don't want to go back. But we are facing a crisis, and this is about going forwards in a way that will work, that is resilient, sustainable and fair. And protecting nature and biodiversity is as important as tackling climate change. It feels more complicated and messier. Unlike replacing fossil fuel use with renewable energy, we can't just substitute one source of biodiversity for another. But we can all rebug.

This book came to me as I started to imagine everyone rebugging, in whatever way they can. Step by step, wildflower by wildflower, verge by verge, meal by meal. This is already happening. I see more interest, more actions and more recognition of the role of nature and the bugs everywhere I go.

— CHAPTER 9 —

# You don't have to rebug alone

You may choose to go it alone and use these resources for advice and ideas on rebugging. That's great. Bugs will make great companions on this journey in my opinion. But you may also not want to do it all alone. There are many organisations and hundreds of tools you can turn to for help in rebugging the planet.

This chapter begins with a basic guide on change making – the political side of rebugging that is lobbying and campaigning. Following that is a directory of many of the organisations you can join and to find resources you need.

I have grouped them into rough themes, including: food; consumer groups; invertebrate and wider environmental, wildlife and habitat organisations; and those that protect specific sites that you can visit and help conserve. Inevitably, many groups undertake several activities, such as those providing information and practical tips as well as helping you to campaign locally and nationally.

Many conservation organisations also have their own reserves and sites you can visit and help to protect. I have included global organisations, including those in the UK and US.

There is no right or wrong way to do any of this, from building a bug-friendly garden plot to writing to the prime minister or president. I hope you find what you are looking for and get to rebug with pleasure and success.

## How to start lobbying for change

As the previous chapters show, much of what needs to change is not just down to you but requires government, industry and even international action. This means telling them what you want to see changed and campaigning to make them act.

At its simplest, lobbying is about letting those who make decisions hear what you have to say and what your evidence is. This may be your local councillor or council staff, civil servants, your local representative in parliament or a minister, prime minister or a president. Never think you can't or that actions won't make a difference. You can write, arrange to see them, invite them to an event you may be running or join an organised event.

Taking it to the next level of lobbying will be about influencing them to make the decisions you want them to make, the changes to want to see. This may be about the local council cutting herbicide use on local road verges and parks, or getting central government to pay farmers differently, so invertebrates are better protected.

### *First steps as a lobbyist*

You need to decide exactly what you hope to achieve by lobbying, who can help, what motivates them, what could catch their eye. You need to be clear about your facts. Lobbying is not about persuading people to do something by the force of your personality (that can both help or hinder). It is about giving the right people the right information at the right time in the right way.

In deciding whom you need to speak to, you need to ask yourself:

1. Who has the power to take the decision I want?
2. What will interest them?
3. Who else would they listen to?
4. How can I make it easy for them to agree with me?

Lobbying central government is not easy as they are harder to reach than local. Going through your parliamentary representative can really help as they have access to ministers. Going in greater numbers can help. Form or join an effective pressure group. Get the group behind you. Try to make headlines in your local media about the issue and link to local stories, visuals, locations and communities as that will boost the likelihood of getting local media coverage.

Using the media is always helpful as the government does respond to national headlines and officials certainly respond to local media.

When you write to your parliamentary representative, be aware you may need to write again and be persistent. They may have local constituency offices and regular open meetings, called surgeries, so you can write and meet them at these sessions. Check with the constituency office or the representatives' office whether an appointment is necessary before you go.

Try to work out beforehand how they may respond, what their interests are, and what research and numbers of constituents you may need to present, to convince them of your case. Reading about what they have supported before can be useful in understanding what they have shown interest in.

Going directly to the minister is harder to do and may be easier as part of a bigger group as ministers tend not to meet with ordinary citizens. The minister should give a personal reply, though the letter will still be drafted by civil servants.

Writing as a representative of a national organisation and institution can help. Other tools include finding out if the minister has said anything or shown interest in your issue. And if you have managed to generate media coverage, this will open doors. Getting politicians interested is an art form, but you can also simply be lucky with the right campaign at the right time.

There are also a whole range of people who can influence a minister including their staff, researchers, people in their political party headquarters, members of their party (you may know some even if you don't belong) and local councillors within their constituency.

Finally, some key things to remember whomever you are wanting to influence:

- Have all the facts to hand and a good story to tell if possible. Think of visuals – handing over a big petition or a tree to plant, having someone dressed as a bee – if a photo opportunity is needed.
- Practice with friends beforehand how you will explain the change you feel is necessary. Get your friends to think about what questions may come back at you. A bit of role play can be hugely valuable.
- Leave something with them summarising your request and your evidence, but always follow up with a letter to answer questions that arose and repeat your asks.
- Keep bugging them. Always. For the bugs.

Now here's a list of the organisations that will support your rebugging efforts.

*Gardening, farming and food*

Back from the Brink – www.naturebftb.co.uk

The aim of Back from the Brink is to save twenty species from extinction and benefit over two hundred more through nineteen projects that span England; from the tip of Cornwall to Northumberland.

Biodynamic Association – www.biodynamic.org.uk

A charitable organisation founded in the UK in 1929 to promote a form of agroecological and nature-focused farming and gardening called biodynamic.

Capital Growth – www.capitalgrowth.org

Capital Growth, a project of Sustain, supports people to grow food in London, whether at home, on allotments or as part of a community group. It provides help to people who grow their own food in London, including discounted training, networking events, support with growing to sell and discounts on equipment. It's free to join and they are always on the lookout for volunteers and trainers.

Ethical Consumer – www.ethicalconsumer.org

Ethical Consumer is an independent, not-for-profit, cooperative with open membership, which provides all the tools and resources needed to make green, fair and healthier choices at the checkout.

Farming and Wildlife Advisory Group – www.fwag.org.uk/about-fwag

A group providing practical and independent environmental and conservation advice to the farming community.

Garden Organic – www.gardenorganic.org.uk
Helping gardeners and growers use organic tools in windowsills, gardens, community spaces and allotments.

Incredible Edible Network – www.incredibleedible.org.uk
The Incredible Edible Network was set up in 2012 in response to the huge popularity of the original group in Todmorden in the UK and the huge number of enquiries from those who wanted to do something similar – creating kind, confident and connected communities through food growing.

Innovate Farmers – www.innovativefarmers.org
A network of farmers of all shapes and sizes doing farm research on the way they want to do it and with a great focus on researching viable farming using sustainable, wildlife friendly systems.

LEAF (Linking Environment and Farming) – www.leafuk.org
LEAF works to inspire and enable sustainable farming in the UK and globally, and has a certification scheme and training programme.

Marine Conservation Society – www.mcsuk.org
The Marine Conservation Society is the UK's leading charity for the protection of our seas, shores and wildlife.

Marine Stewardship Council – www.msc.org/home
The MSC uses its ecolabel and fishery certification programme to drive uptake of better fishing practices by providing market rewards, and so contributes to the health of the world's oceans.

Nature Friendly Farming Network – www.nffn.org.uk
The Nature Friendly Farming Network is led by farmers with a passion for sustainable farming and nature. They work to unite farmers across the UK who farm for

wildlife and the environment, and lobby for changes in policy, including on public support for good farming.

OF&G (Organic Farmers and Growers) – www.ofgorganic.org
OF&G certify more than half of UK organic land and provide support, information and licencing to Britain's top organic food businesses.

Pasture for Life – www.pastureforlife.org
The Pasture-Fed Livestock Association was created to promote the unique quality of livestock produce raised exclusively on pasture, and the wider environmental and animal welfare benefits that such pastured livestock systems provide.

Permaculture Association – www.permaculture.org.uk
The Permaculture Association help its members and others in the permaculture network to promote permaculture ethics and principles for sustainable food systems, and are part of a worldwide permaculture community.

Soil Association – www.soilassociation.org
The Soil Association is one of the leading membership charities campaigning for healthy, humane and sustainable food, farming and land use in the UK. It also runs a subsidiary, Soil Association Certification Limited, to certify organic foods.

SRI Services (Sustainable and Responsible Investment) – www.sriservices.co.uk
An organisation supporting sustainable, responsible and ethical investment.

Wildlife Gardening Forum – www.wlgf.org
The Wildlife Gardening Forum want more people to discover how important gardens can be for wildlife such as

bugs, and to get enjoyment, learning, playing and health benefits from great gardens for nature.

The Woodland Trust – www.woodlandtrust.org.uk

The Woodland Trust works to achieve a UK far richer in woods and trees, to create habitats for wildlife like bugs, capture carbon and build better places to live and spend time in.

WorldWide Opportunities on Organic Farms (WWOOF) – www.wwoof.org.uk

WWOOF UK is part of a worldwide movement linking people who want some experience growing organic food with organic farmers and growers who need them, to build a great community of growers knowing how to grow or farm with wildlife in mind.

*Invertebrate-focused organisations*

Amphibian and Reptile Conservation (ARC) – www.arc-trust.org

ARC works to protect amphibians and reptiles, and the habitats they need to live, feed and breed in. They run sites and manage volunteers to help protect these vital oases for water-needing species and the bugs they depend on.

Bat Conservation Trust – www.bats.org.uk

Clearly bats need bugs and the Bat Conservation Trust, which supports local bat groups across the UK, has over six thousand members and runs national and local projects which support bug habitats as well as bats!

Bee Friendly Monmouthshire – www.beefriendlymonmouthshire.org/about

The local organisation working to ensure that Monmouthshire, UK is pollinator friendly with lots of ideas and advice.

Buglife – www.buglife.org.uk
Buglife is a critical organisation devoted to the conservation of all invertebrates and works to save Britain's bugs, from bees to beetles and worms to woodlice.

Bumblebee Conservation Trust – www.bumblebee conservation.org
The Bumblebee Conservation Trust was established to address major problems for bumblebees and work to ensure bumblebees are thriving and valued.

Butterfly Conservation – www.butterfly-conservation.org
Butterfly Conservation works through conservation projects and campaigns to conserve butterflies and moths – a vital part of our wildlife heritage and great indicators of the health of our environment.

National Insect Week – www.nationalinsectweek.co.uk
National Insect Week was set up by the Royal Entomological Society to encourage people of all ages to learn more about insects, with many partner organisations and events.

Royal Entomological Society – www.royensoc.co.uk
The Royal Entomological Society was formed to promote and develop insect science and it supports international collaboration, research and has several top publications. It also runs National Insect Week.

The Amateur Entomologists' Society – www.amentsoc.org
The Amateur Entomologists' Society is a charity, founded in 1935, by volunteers with an interest in insects. It promotes the study of entomology, especially among amateurs and the younger generation.

Zoological Society London (ZSL) – www.zsl.org
The ZSL is an international conservation charity that works

to promote wildlife conservation via science, global field conservation and engagement and public interest through visitors to its two zoos.

*Wildlife organisations who provide advice, actions and campaign*

Amphibian and Reptile Conservation (ARC) – www.arc-trust.org

ARC works to conserve amphibians and reptiles, and the habitats (and bugs) they depend on, to protect them for future generations.

BirdLife International – www.birdlife.org

BirdLife International is the global network of all the national conservation organisations with a focus on bird life but advocating and carrying out priority conservation actions, which affect all wildlife including bugs, of course.

British Ecological Society – www.britishecologicalsociety.org

The British Ecological Society was established in 1913 to promote the science of ecology and has members around the world to advance ecological science.

Campaign for National Parks (CNP) – www.cnp.org.uk

The CNP is dedicated to campaigning to protect and promote all of the national parks of England and Wales.

CHEM Trust – www.chemtrust.org

CHEM Trust works on projects to stop man-made chemicals from causing long-term damage to humans or wildlife, including the bugs, and promoting alternative approaches.

ClientEarth – www.clientearth.org
ClientEarth uses the law to make systemic changes to protect the environment because it's often the most powerful tool available.

The Conservation Volunteers – www.tcv.org.uk
The Conservation Volunteers is a UK community volunteering charity, which creates healthier and happier communities for everyone by encouraging volunteering in conservation.

CPRE: the countryside charity – www.cpre.org.uk
CPRE, the countryside charity, has local groups in every county, and advocates nationwide for a countryside with sustainable, healthy communities and that is available to everyone.

Environmental Investigation Agency – www.eia-international.org
The EIA undertake major investigations into environmental crime and wildlife abuse globally.

Fauna and Flora International – www.fauna-flora.org
Flora and Fauna International is the world's oldest international wildlife conservation organisation and focuses on protecting biodiversity globally.

Field Studies Council – www.field-studies-council.org
The Field Studies Council runs courses and field trips and offers opportunities for people to learn and engage with the outdoors for their benefit.

Friends of the Earth – England and Wales: www.friendsoftheearth.uk, Scotland: www.foe.scot
Friends of the Earth is a grassroots environmental campaigning organisation, which works from the ground up, with

local groups, campaigners and lawyers to push for change in environmental policy and practice.

Global Canopy Programme – www.globalcanopy.org
The Global Canopy Programme focuses on the production, trade and financing of commodities such as soy, beef and palm oil that are responsible for the majority of deforestation worldwide.

Greenpeace – www.greenpeace.org.uk
Greenpeace is an international movement of people who are passionate about defending the natural world from destruction.

Groundwork – www.groundwork.org.uk
Groundwork is a UK federation of charities working locally and nationally to transform lives in the UK's most disadvantaged communities by helping them to get involved in local projects, improving green space.

National Trust – www.nationaltrust.org.uk
The National Trust looks after the nation's coastline, historic sites, countryside and green spaces, and works to ensure everyone benefits.

People's Trust for Endangered Species – www.ptes.org
The People's Trust for Endangered Species was set up in 1977 to focus campaigning on the most endangered species.

Pesticide Action Network – www.pan-uk.org
PAN-UK is the only UK charity focused on tackling the problems caused by pesticides and promoting safe and sustainable alternatives in agriculture, urban areas, homes and gardens. As such they do so much for the bugs! They campaign for change in policy and practices at home and overseas, coordinate projects which help

smallholder farming communities escape ill-health and poverty caused by pesticides, and share scientific and technical expertise.

Plantlife – www.plantlife.org.uk
Plantlife is a British conservation charity set up to save threatened wild flowers, plants and fungi. They run nature reserves across England, Scotland and Wales, and work with landowners, businesses, conservation organisations, community groups and governments, to save the rarest flora and ensure familiar flowers and plants can thrive.

Rewilding Britain – www.rewildingbritain.org.uk
Rewilding Britain want rewilding to flourish across Britain to tackle the climate emergency and extinction crisis, reconnect people with the natural world, and help individuals and communities thrive with new opportunities.

Rewilding Europe – www.rewildingeurope.com
Rewildling Europe want to make Europe a wilder place, with more space for wild nature, wildlife and natural processes. They see rewilding as a critical opportunity to rebuild wild areas, help people to enjoy wildlife and earn a fair living from the wild.

RSPB – www.rspb.org.uk
RSPB is the largest nature conservation charity in the UK, running successful conservation projects, lobbying and campaigning for wildlife, and especially the birds and the wildlife they rely on.

Scottish Wildlife Trust – www.scottishwildlifetrust.org.uk
The Scottish Wildlife Trust is a leading charity dedicated to nature conservation in Scotland.

Surfers Against Sewage (SAS) – www.sas.org.uk
SAS is a UK grassroots movement dedicated to the protection of oceans, waves, beaches and wildlife.

Trees For Life – www.treesforlife.org.uk
Trees For Life works to deliver an ambitious vision of a revitalised wild forest in the Highlands of Scotland, providing space for wildlife there including the wood ants.

Wildfowl and Wetland Trust (WWT) – www.wwt.org.uk
The WWT is the UK's leading wetland conservation charity saving threatened wetland wildlife, and provides science and experts in wetland management and creation.

Wild Justice – www.wildjustice.org.uk
Wild Justice has been set up to use legal tools and work to change the law to protect the wildlife – as it can't take legal cases in its own.

The Wildlife Trusts – www.wildlifetrusts.org
The Wildlife Trusts has trusts across the UK working with volunteers, communities and landowners to make the world wilder and make nature part of life, for everyone.

Wildlife Trusts Wales – www.wtwales.org
Wildlife Trusts Wales believes that wildlife and natural processes need to have space to thrive, beyond designated nature reserves and other protected sites.

WWF – www.wwf.org.uk
WWF is a leading independent conservation organisation.

*US conservation organisations*

Audubon – www.audubon.org

The National Audubon Society works to protects birds and the places they need throughout the Americas using science, advocacy, education and on-the-ground conservation.

Conservation International – www.conservation.org

Conservation International works to protect nature through cutting-edge science, innovative policy and global reach; they empower people to protect the nature that we rely on for food, fresh water and livelihoods.

Sierra Club – www.sierraclub.org

The Sierra Club is an influential grassroots environmental US organisation with 3.8 million members and supporters working to defend everyone's right to a healthy world.

The Jane Goodall Institute – www.janegoodall.org

The Jane Goodall Institute promotes understanding and protection of great apes and their habitat and builds on the legacy of Dr. Jane Goodall, its founder, to inspire individual action by young people of all ages to help the environment and wildlife.

The National Wildlife Federation (NWF) – www.nwf.org

The NWF protects natural resources for hunters, anglers, hikers, birders, wildlife watchers, boaters, climbers, campers, cyclists, gardeners, farmers, forest stewards and other outdoor enthusiasts.

The Nature Conservancy – www.nature.org

The Nature Conservancy is a global environmental charity working to create a world where people and nature can thrive.

The Wildlife Conservation Society (WCS) – www.wcs.org
The WCS goal is to conserve the world's largest wild places in fourteen priority places, home to over half of the planet's biodiversity.

The World Wildlife Fund (WWF) – www.worldwildlife.org
WWF's work has evolved from purely saving species and landscapes to trying to address the larger global threats and drivers that cause damage.

The Xerces Society for Invertebrate Conservation – www.xerces.org
The Xerces Society for Invertebrate Conservation is an international organisation that works for the conservation of invertebrates and their habitats. Its name (which is pronounced Zer-sees, or /ˈzɚˌsiz/) comes from the now-extinct Xerces blue butterfly (*Glaucopsyche xerces*), the first butterfly known to go extinct in North America as a result of human activities.

### *Local organisations*

You will find local and community groups in your local newspapers, directories and via local websites. Look for conservation and wildlife trusts, local volunteering bodies and ask your local authority such as your district, county, borough or parish council for a list or look on their websites for conservation, green activities and volunteering opportunities. Local wildlife and conservation trusts will be invaluable. Local farmers and landowners may already be involved with local organisations identifying and tracking wildlife on the farm.

## Useful guides on how to campaign and influence people

Crucially, you don't have to campaign alone. Asking your family and friends to join you will be extremely valuable. Your local community may also help. Several organisations, like Citizens UK, are dedicated to help support community organising, which is based on the principle that when people work together they have the power to change their neighbourhoods, cities and ultimately the country for the better.

The following sites have useful tools and ideas for how to campaign, case studies you can learn from, lobby ideas, how to write letters, start a community action, petitions, organising and more:

https://www.citizensuk.org – Citizens UK organises communities to act together for power, social justice and the common good. We are the home of community organising in the UK.

https://publicinterest.org.uk – Public Interest works with civil society to develop stories and strategies for a more equal, green and democratic society.

https://mobilisationlab.org – MobLab equips advocacy campaigners and their organisations to win in the networked age with transformative, participatory and creative approaches to social change.

https://trainings.350.org – training and tools with the 350 organisation on how to make change happen.

http://www.pan-uk.org/pesticide-free – the tools and information you need to run a pesticide-free town campaign.

www.change.org – tools to set up a petition, start campaigns, mobilise supporters and work with decision makers to drive solutions.

https://secure.avaaz.org – a global organisation empowering millions of people from all walks of life to take action on pressing global, regional and national issues, from corruption and poverty to conflict and climate change.

Some useful resources on how to find your MP, council, and tools in the UK:

https://www.gov.uk/find-local-council – who and how to contact your local council

https://www.gov.uk/make-a-freedom-of-information-request – how to make a Freedom of Information request, for instance to your local authority to find out what chemicals they use.

https://members.parliament.uk/FindYourMP – how to find your MP, what they have done and said, and how to contact them.

https://www.parliament.uk/get-involved/contact-your-mp – the official site for UK parliament.

To locate your senator using Congress.gov, visit the Congress.gov homepage: https://www.congress.gov. To find your representative, you will first have to determine what congressional district you live in. To do this, visit the House of Representatives' 'Find Your Representative': https://www.house.gov/representatives/find-your-representative.

https://www.mysociety.org/wehelpyou/who-is-your-mp – mySociety is a not-for-profit social enterprise, based in the UK but working with partners internationally. They build and share digital technologies that help people be active citizens, across the three areas of democracy, transparency and community.

# ACKNOWLEDGEMENTS

This book was inspired by so many people I cannot begin to thank them. The many conservationists and researchers who I have drawn on and the writers and campaigners who have inspired me. I hope I have referenced adequately, and all errors and omissions are obviously mine alone.

I am grateful to Jon Rae, Michael Metivier and Rose Baldwin at Chelsea Green for giving me this great chance to realise a project I never thought would see daylight. Muna Reyal at Chelsea Green has been a wonderful editor and made what would have been far too indigestible, a good read. I would also like to thank Ned Page for his wonderful illustrations, and Tom Rice-Hird for helping out, and all who know me for giving great advice and looking out for me, especially all the great team at Sustain. I wrote a third of this in the Covid-19 lockdown and am eternally grateful to my family for their patience and silence when needed.

# NOTES

## Foreword

1. Rodolfo Dirzo et al., 'Defaunation in the Anthropocene', *Science* 345, no. 6195 (July 2014): 401–06, http://doi.org/10.1126/science.1251817.
2. Howard Dryden and Diane Duncan, 'Plastic and Chemicals Toxic to Plankton Will Accelerate Ocean Acidification Which Could Devastate Humanity in 25 Years Unless We Stop the Pollution', *Environmental Science eJournal* 1, no. 28 (June 2021), http://dx.doi.org/10.2139/ssrn.3860950.
3. Caspar A. Hallmann et al., 'More Than 75 Percent Decline Over 27 Years in Total Flying Insect Biomass in Protected Areas', *PLoS ONE* 12, no. 10 (October 2017): e0185809, https://doi.org/10.1371/journal.pone.0185809; Hallmann et al., 'Declining Abundance of Beetles, Moths and Caddisflies in the Netherlands', *Insect Conservation and Diversity* 13, no. 2 (March 2020): 127–39, https://doi.org/10.1111/icad.12377.

## Introduction

1. Francisco Sánchez-Bayo and Kris A. G. Wyckhuys, 'Worldwide Decline of the Entomofauna: A Review of Its Drivers', *Biological Conservation* 232 (April 2019): 8–27, https://doi.org/10.1016/j.biocon.2019.01.020.
2. Simon Leather 'Insectageddon" – Bigger Headlines, More Hype, but Where's the Funding?' *Don't Forget The Roundabouts* blog, 15 February 2019, https://simonleather.wordpress.com/2019/02/15/insectageddon-bigger-headlines-more-hype-but-wheres-the-funding/.
3. Dave Goulson, *Insect Declines and Why They Matter*, Report Commissioned by the South West Wildlife Trusts, https://www.flipsnack.com/devonwildlifetrust/insect-declines/full-view.html.
4. Ollerton, J., Winfree, R. and Tarrant, S., 'How Many Flowering Plants are Pollinated by Animals?', *Oikos*, 120 (February 2011): 321–326,

https://doi.org/10.1111/j.1600-0706.2010.18644.xhttps://online library.wiley.com/doi/full/10.1111/j.1600-0706.2010.18644.x.

5. Dave Goulson, 'Are Robotic Bees the Future?', *Dave's Blog*, University of Sussex, 16 October, 2018, http://www.sussex.ac.uk/lifesci/goulson lab/blog/robotic-bees.
6. Arno Thielens et al., 'Radio-Frequency Electromagnetic Field Exposure of Western Honey Bees', *Science Reports* 10, no. 461 (2020), https://doi.org/10.1038/s41598-019-56948-0.

## Chapter 1. Rebugging our attitudes

1. Michael J. Samways et al., 'Solutions for Humanity on How to Conserve Insects', *Biological Conservation* 242 (February 2020): 108427, https://doi.org/10.1016/j.biocon.2020.108427.
2. Patrick Greenfield, '"Sweet City": The Costa Rica Suburb that Gave Citizenship to Bees, Plants and Trees', *Guardian*, 29 April 2020, https://www.theguardian.com/environment/2020/apr/29/sweet -city-the-costa-rica-suburb-that-gave-citizenship-to-bees-plants -and-trees-aoe.
3. Patrick Greenfield, '"Sweet City": The Costa Rica Suburb that Gave Citizenship to Bees, Plants and Trees', *Guardian*, 29 April 2020, https://www.theguardian.com/environment/2020/apr/29/sweet -city-the-costa-rica-suburb-that-gave-citizenship-to-bees-plants -and-trees-aoe.
4. Municipality of Curridabat, 'Curridabat: Sweet City: A City Modelling Approach Based Pollination', *Curridabat Sweet City Magazine*, https://static1.squarespace.com/static/5bbd32d6e66669016 a6af7e2/t/5c757759e2c4835d3cbc174f/1551202139913/Currida bat_Sweet_City_Magazine.pdf.
5. Desmond Pugh, 'Monmouth Can Become World's First Bee Town', *Monmouthshire Beacon*, 11 October 2019, http://www.monmouthshire beacon.co.uk/article.cfm?id=117564.
6. Qijue Wang and Hannes C. Schniepp, 'Strength of Recluse Spider's Silk Originates from Nanofibrils', *ACS Macro Letters* 7, no. 11 (October 2018): 1364–70, https://doi.org/10.1021/acsmacrolett.8b00678.
7. Jesus Rivera et al., 'Toughening Mechanisms of the Elytra of the Diabolical Ironclad Beetle', *Nature* 586 (October 2020): 543–48, https://doi.org/10.1038/s41586-020-2813-8.
8. Hao Liu et al., 'Biomechanics and Biomimetics in Insect-Inspired Flight Systems', *Philosophical Transactions of the Royal Society B* 371,

no. 1704 (September 2016), https://doi.org/10.1098/rstb.2015.0390.

9. Pan Liu et al., 'Flies Land Upside Down on a Ceiling Using Rapid Visually Mediated Rotational Maneuvers', *Science Advances* 5, no. 10 (October 2019): eaax1877, https://doi.org/10.1126/sciadv.aax1877.
10. Bjorn Carey, 'Stanford Researchers Discover the "Anternet"', Stanford News, 24 August 2012, https://news.stanford.edu/news/2012/august/ants-mimic-internet-082312.html.
11. Yves Basset et al., 'Arthropod Diversity in a Tropical Forest', *Science* 338, no. 6113 (December 2012): 1481–84, https://doi.org/10.1126/science.1226727.
12. Nigel E. Stork and Jan C. Habel, 'Can Biodiversity Hotspots Protect More Than Tropical Forest Plants and Vertebrates?', *Journal of Biogeography* 41, no. 3 (March 2014): 421–28, https://doi.org/10.1111/jbi.12223.
13. Louisa Casson, 'Protecting Nature Means Protecting Ourselves', Greenpeace, 22 May 2020, https://www.greenpeace.org/international/story/43423/protecting-nature-means-protecting-ourselves/.
14. Erica L. Morley and Daniel Robert, 'Electric Fields Elicit Ballooning in Spiders', *Current Biology* 28, no. 14 (July 2018): 2324–30, https://doi.org/10.1016/j.cub.2018.05.057.
15. John E. Losey and Mace Vaughan, 'The Economic Value of Ecological Services Provided by Insects', *Bioscience* 56, no. 4 (April 2006): 311–23, https://doi.org/10.1641/0006-3568(2006)56[311:TEVOES]2.0.CO;2.
16. David Kleijn et al., 'Delivery of Crop Pollination Services is an Insufficient Argument for Wild Pollinator Conservation', *Nature Communications* 6 (2015): 7414, https://doi.org/10.1038/ncomms8414.
17. Sarah Knapton, 'Bees Contribute More to British Economy Than Royal Family', *Telegraph*, 17 June 2015, https://www.telegraph.co.uk/news/earth/wildlife/11679210/Bees-contribute-more-to-British-economy-than-Royal-Family.html.
18. Simon G. Potts, Vera L. Imperatriz-Fonseca, and Hien T. Ngo, eds., 'Assessment Report on Pollinators, Pollination and Food Production', Intergovernmental Science-Policy Platform on Biodiversity and Ecosystem Services (7 December 2016), https://doi.org/10.5281/zenodo.3402856.

## Chapter 2. What bugs do for us

1. Kris A.G. Wyckhuys et al., 'Biological Control of an Agricultural Pest Protects Tropical Forests', *Communications Biology* 2, no. 10 (January 2019), https://doi.org/10.1038/s42003-018-0257-6.
2. M. Subrahmanyam, 'Topical Application of Honey in Treatment of Burns', *Br J Surg*, 78 (4), (April 1991):497–98, https://doi: 10.1002/bjs.1800780435. PMID: 2032114.
3. 'Edible Insects for Animal Feed', Persistence Market Research (April 2020), https://www.persistencemarketresearch.com/market-research/edible-insects-for-animal-feed-market.asp.

## Chapter 3. Rewilding by rebugging

1. Isabella Tree, *Wilding: The Return of Nature to a British Farm* (London: Picador, 2018).
2. Isabella Tree, *Wilding: The Return of Nature to a British Farm* (London: Picador, 2018).
3. National Wood Ant Steering Group, 'UK Wood Ants', The James Hutton Institute, https://www.woodants.org.uk/species/significance.
4. Jenni A. Stockan et al., 'Wood Ants and Their Interaction with Other Organisms', *Wood Ant Ecology and Conservation* 8 (June 2016): 177–206, https://doi.org/10.1017/CBO9781107261402.009.
5. Douglas H. Chadwick, 'Keystone Species: How Predators Create Abundance and Stability', *Mother Earth News*, 1 June 2011, https://www.motherearthnews.com/nature-and-environment/wildlife/keystone-species-zm0z11zrog#ixzz1clbGyAwq.
6. Neus Rodríguez-Gasol et al., 'The Contribution of Surrounding Margins in the Promotion of Natural Enemies in Mediterranean Apple Orchards', *Insects* 10, no. 5 (2019), https://doi.org/10.3390/insects10050148.
7. Brett R. Blaauw and Rufus Isaacs, 'Flower Plantings Increase Wild Bee Abundance and the Pollination Services Provided to a Pollination-Dependent Crop', *Journal of Applied Ecology* (March 2014), https://doi.org/10.1111/1365-2664.12257.
8. 'Wood Ants (*Formica aquilonia, Formica lugubris*) Play a Very Important Role in the Caledonian Forest Ecosystem', Trees for Life, https://treesforlife.org.uk/into-the-forest/trees-plants-animals/insects-2/.
9. Vicki Hird, 'We Have an Agriculture Act – But Let's Not Relax Now', *Sustainable Farming Policy* (blog), Sustain, 11 November 2020, https://www.sustainweb.org/blogs/nov20-new-agriculture-act2020/.

10. Timothy C. Winegard, *The Mosquito: A Human History of Our Deadliest Predator* (New York: Dutton, 2019).

## Chapter 4. Parks and recreation: rebugging your world

1. Luis Mata et al., 'Conserving Herbivorous and Predatory Insects in Urban Green Spaces', *Scientific Reports* 7 (January 2017): 40970, https://doi.org/10.1038/srep40970.
2. 'Pesticide-Free Towns Campaign', Pesticide Action Network UK, https://www.pan-uk.org/pesticide-free/.
3. 'Breaking New Ground with Eco Drive to Bring the Country's Verges to Life', Highways England, 2 December 2020, https://www.gov.uk/government/news/breaking-new-ground-with-eco-drive-to-bring-the-countrys-verges-to-life.
4. Capital Growth, 'London Grows Wild: A Guide to Wildlife-friendly Food Growing', Sustain, 19 September 2016, https://www.sustainweb.org/publications/london_grows_wild.
5. 'Simple Things You Can Do to Help Wildlife,' Wildlife Trusts, https://www.wildlifetrusts.org/actions.
6. Robert A. Hammond and Malcolm D. Hudson, 'Environmental Management of UK Golf Courses for Biodiversity – Attitudes and Actions', *Landscape and Urban Planning* 83, no. 2–3 (November 2007): 127–36, https://doi.org/10.1016/j.landurbplan.2007.03.004.
7. Mata, 'Conserving Herbivorous and Predatory Insects', 40970.
8. Witney Town Council, 'Witney Tiny Forest', (March 2020): http://www.witney-tc.gov.uk/witneys-tiny-forest/; 'UK's First-Ever Tiny Forest Seeks to Deliver Big Benefits for People and the Environment', Earthwatch, https://earthwatch.org.uk/component/k2/tiny-forest.

## Chapter 5. The bigger bug challenges

1. Graham A. Montgomery et al., 'Is the Insect Apocalypse Upon Us? How to Find Out', *Biological Conservation* 241 (January 2020): 108327, https://doi.org/10.1016/j.biocon.2019.108327.
2. Christopher A. Halsch et al., 'Pesticide Contamination of Milkweeds Across the Agricultural, Urban, and Open Spaces of Low-Elevation Northern California', *Frontiers in Ecology and Evolution* (2020): https://doi.org/10.3389/fevo.2020.00162.
3. Swantje Grabener et al., 'Changes in Phenology and Abundance of Suction-Trapped Diptera from a Farmland Site in the UK over Four

Decades', *Ecological Entomology* 45, no. 5 (October 2020): 1215–19, https://doi.org/10.1111/een.12873.

4. A. Atkinson et al., 'A Re-appraisal of the Total Biomass and Annual Production of Antarctic Krill', *Deep-Sea Research Part I* 56, no. 5 (May 2009): 727–40, https://doi.org/10.1016/j.dsr.2008.12.007.
5. Daniel G. Boyce et al., 'Global Phytoplankton Decline over the Past Century', *Nature* 466 (2010): 591–96, https://doi.org/10.1038/nature 09268.
6. Peter Soroye et al., 'Climate Change Contributes to Widespread Declines among Bumble Bees across Continents,' *Science* 367, no. 6478 (February 2020): 685–88, https://doi.org/10.1126/science .aax8591.
7. Helen R.P. Phillips et al., 'Global Distribution of Earthworm Diversity', *Science* 366, no. 6464 (October 2019): 480–85, https://doi.org /10.1126/science.aax4851.
8. Jean M. Holley and Nigel R. Andrew, 'Experimental Warming Disrupts Reproduction and Dung Burial by a Ball-Rolling Dung Beetle', *Ecological Entomology* 44, no. 2 (April 2019): 206–16, https://doi.org /10.1111/een.12694.
9. Anna-Christin Joel et al., 'Biomimetic Combs as Antiadhesive Tools to Manipulate Nanofibers', *ACS Applied Nano Materials* 3, no. 4 (2020): 3395–3401, https://doi.org/10.1021/acsanm.0c00130.
10. Alexander G. Little et al., 'Population Differences in Aggression Are Shaped by Tropical Cyclone-Induced Selection', *Nature Ecology & Evolution* 3 (2019): 1294–97, https://doi.org/10.1038/s41559-019-0951-x.
11. Amanda M. Koltz and Justin P. Wright, 'Impacts of Female Body Size on Cannibalism and Juvenile Abundance in a Dominant Arctic Spider', *Journal of Animal Ecology* 89, no. 8 (August 2020): 1788–98, https://doi.org/10.1111/1365-2656.13230.
12. Sandra Díaz et al., 'Summary for Policymakers of the Global Assessment Report on Biodiversity and Ecosystem Services', Intergovernmental Science-Policy Platform on Biodiversity and Ecosystem Services, 25 November 2019, https://doi.org/10.5281 /zenodo.3553579.
13. Stephen J. Martin et al., 'A Vast 4,000-Year-Old Spatial Pattern of Termite Mounds', *Current Biology* 28, no. 22 (November 2018): 1292–93, https://doi.org/10.1016/j.cub.2018.09.061.
14. Katie Burton, 'The Story Behind Brazil's 200 Million Termite Mounds', *Geographical* (2019), https://geographical.co.uk/nature

/wildlife/item/3053-the-story-behind-brazil-s-200-million-termite-mounds.

15. Penelope A. Hancock et al., 'Mapping Trends in Insecticide Resistance Phenotypes in African Malaria Vectors', *PLOS Biology* 18, no. 6 (2020): e3000633, https://doi.org/10.1371/journal.pbio.3000633.
16. Peter Dizikes, 'Out of Thick Air', *MIT News*, 21 April 2011, http://news.mit.edu/2011/fog-harvesting-0421.
17. Charlotte Bruce-White and Matt Shardlow, 'A Review of the Impact of Artificial Light on Invertebrates', Buglife (2011), https://cdn.buglife.org.uk/2019/08/A-Review-of-the-Impact-of-Artificial-Light-on-Invertebrates-docx_0.pdf.
18. Erin P. Walsh et al., 'Noise Affects Resource Assessment in an Invertebrate', *Biology Letters* 13, no. 4 (April 2017), https://doi.org/10.1098/rsbl.2017.0098.
19. Hansjoerg P. Kunc and Rouven Schmidt, 'The Effects of Anthropogenic Noise on Animals: a Meta-analysis', *Biology Letters* 15, no. 11 (November 2019), http://doi.org/10.1098/rsbl.2019.0649.
20. Arno Thielens et al., 'Exposure of Insects to Radio-Frequency Electromagnetic Fields from 2 to 120 GHz', *Science Reports* 8, no. 3924 (2018), https://doi.org/10.1038/s41598-018-22271-3.
21. Arno Thielens et al., 'Radio-Frequency Electromagnetic Field Exposure of Western Honey Bees', *Science Reports* 10, no. 461 (2020), https://doi.org/10.1038/s41598-019-56948-0.

## Chapter 6. Why our farming, food and shopping need bugs

1. Manu E. Saunders, 'Resource Connectivity for Beneficial Insects in Landscapes Dominated by Monoculture Tree Crop Plantations', *International Journal of Agricultural Sustainability* 14, no. 1 (2016): 82–99, https://doi.org/10.1080/14735903.2015.1025496.
2. Amanda E. Martin et al., 'Effects of Farmland Heterogeneity on Biodiversity Are Similar to – or Even Larger Than – the Effects of Farming Practices', *Agriculture, Ecosystems & Environment* 288 (February 2020): 106698, https://doi.org/10.1016/j.agee.2019.106698.
3. Foteini G. Pashalidou et al., 'Bumble Bees Damage Plant Leaves and Accelerate Flower Production When Pollen Is Scarce', *Science* 368, no. 6493 (May 2020): 881– 84, https://doi.org/10.1126/science.aay0496.
4. Gary D. Powney et al., 'Widespread Losses of Pollinating Insects in Britain', *Nature Communications* 10, no. 1018 (2019), https://doi.org/10.1038/s41467-019-08974-9.

5. 'Saving England's Most Threatened Species from Extinction – Ladybird Spider', Back from the Brink (2020), https://naturebftb.co.uk/the-projects/ladybird-spider/.
6. David Goulson et al., 'Rapid Rise in Toxic Load for Bees Revealed by Analysis of Pesticide Use in Great Britain', *PeerJ* 6, e5255 (2018), https://doi.org/10.7717/peerj.5255.
7. Michael C. Tackenberg et al., 'Neonicotinoids Disrupt Circadian Rhythms and Sleep in Honey Bees', *Scientific Reports* 10, no. 17929 (2020), https://doi.org/10.1038/s41598-020-72041-3.
8. Daniel Schläppi et al., 'Long-Term Effects of Neonicotinoid Insecticides on Ants', *Communications Biology* 3, no. 335 (2020), https://doi.org/10.1038/s42003-020-1066-2.
9. Steven Kragten et al., 'Abundance of Invertebrate Prey for Birds on Organic and Conventional Arable Farms in the Netherlands', *Bird Conservation International* 21, no. 1 (March 2011): 1–11, https://doi.org/10.1017/S0959270910000079.
10. Sean L. Tuck et al., 'Land-Use Intensity and the Effects of Organic Farming on Biodiversity: A Hierarchical Meta-analysis', *Journal of Applied Ecology* 51, no. 3 (2014): 746–55, https://doi.org/10.1111/1365-2664.12219.
11. Friends of the Earth, 'Farming Wheat without Neonicotinoids' and 'Farming Oilseed Rape without Neonicotinoids', Agricology, 2016, https://www.agricology.co.uk/sites/default/files/farming-wheat-without-neonicotinoids-102577.pdf and https://www.agricology.co.uk/resources/farming-oilseed-rape-without-neonicotinoids.
12. Yuta Yamaguchi et al., 'Double-Edged Heat: Honeybee Participation in a Hot Defensive Bee Ball Reduces Life Expectancy with an Increased Likelihood of Engaging in Future Defense', *Behavioral Ecology and Sociobiology* 72, no. 123 (2018), https://doi.org/10.1007/s00265-018-2545-z.
13. The UN Food and Agriculture Organization (FAO) and World Health Organization (WHO) define IPM as follows: 'Integrated pest management means careful consideration of all available plant protection methods and subsequent integration of appropriate measures that discourage the development of populations of harmful organisms and keep the use of plant protection products (pesticides) and other forms of intervention to levels that are economically and ecologically justified and reduce or minimise risks to human health

and the environment. "Integrated pest management" emphasises the growth of a healthy crop with the least possible disruption to agro-ecosystems and encourages natural pest control mechanisms.'

14. Mayara C. Lopes et al., 'Parasitoid Associated with *Liriomyza huidobrensis* (Diptera: Agromyzidae) Outbreaks in Tomato Fields in Brazil', *Agricultural and Forest Entomology* 22 (2020): 224–30, https://doi.org/10.1111/afe.12375.
15. 'Farm to Fork Strategy – for a Fair, Healthy and Environmentally-Friendly Food System' European Commission (2020), https://ec.europa.eu/food/farm2fork_en.
16. IPCC, 'Special Report on Climate Change and Land IPCC, 2019: Summary for Policymakers' in 'Climate Change and Land: an IPCC Special Report on Climate Change, Desertification, Land Degradation, Sustainable Land Management, Food Security, and Greenhouse Gas Fluxes in Terrestrial Ecosystems' (2019), https://www.ipcc.ch/srccl/.
17. Peter Dennis et al., 'The Effects of Livestock Grazing on Foliar Arthropods Associated with Bird Diet in Upland Grasslands of Scotland', *Journal of Applied Ecology* 45 (2008): 279–87, https://doi.org/10.1111/j.1365-2664.2007.01378.x.
18. 'Food Surplus and Waste in the UK – Key Facts Updated January 2020', WRAP, January, 2020, https://wrap.org.uk/sites/files/wrap/Food_surplus_and_waste_in_the_UK_key_facts_Jan_2020.pdf.
19. 'The State of Food and Agriculture', FAO, 2019, http://www.fao.org/3/CA6030EN/CA6030EN.pdf.
20. 'Harnessing the Power of Food Citizenship', Food Ethics Council, 2019, https://www.foodethicscouncil.org/app/uploads/2019/10/Harnessing-the-power-of-food-citizenship_Food_Ethics_Council_Oct-2019.pdf.
21. 'Sustainable Fashion and Textiles', WRAP, accessed 5 July 2020, https://www.wrap.org.uk/content/textiles-overview.
22. 'The TRUE Costs of Cotton: Cotton Production and Water Insecurity', Environmental Justice Foundation, 2012, https://ejfoundation.org/resources/downloads/EJF_Aral_report_cotton_net_ok.pdf.
23. CEPF, 'Ecosystem Profile - Cerrado Biodiversity Hotspot', (2017): 482, https://www.cepf.net/sites/default/files/cerrado-ecosystem-profile-en-updated.pdf.
24. 'Is Cotton Conquering Its Chemical Addiction? Revised June 2018', PAN UK, 2018, https://www.pan-uk.org/cottons_chemical_addiction_updated/.

25. Beverley Henry et al., 'Microfibres from Apparel and Home Textiles: Prospects for Including Microplastics in Environmental Sustainability Assessment', *Science of the Total Environment* 652 (2019): 483–94, https://doi.org/10.1016/j.scitotenv.2018.10.166.
26. Naiara López-Rojo et al., 'Microplastics Have Lethal and Sublethal Effects on Stream Invertebrates and Affect Stream Ecosystem Functioning', *Environmental Pollution* 259 (2019): https://doi.org/10.1016/j.envpol.2019.113898.
27. Penelope K. Lindeque et al., 'Are We Underestimating Microplastic Abundance in the Marine Environment? A Comparison of Microplastic Capture with Nets of Different Mesh-Size', *Environmental Pollution* 265, Part A (2020): 114721, https://doi.org/10.1016/j.envpol.2020.114721.
28. Bas Boots, Connor William Russell, and Dannielle Senga Green 'Effects of Microplastics in Soil Ecosystems: Above and Below Ground', *Environ. Sci. Technol.*, 53 (September 2019): 11496–11506, https://doi.org/10.1021/acs.est.9b03304.
29. Beverley Henry, Kirsi Laitala, Ingun Grimstad Klepp, 'Microfibres from Apparel and Home Textiles: Prospects for Including Microplastics in Environmental Sustainability Assessment', Science ofThe Total Environment 652 (February 2019): 483–94, https://doi.org/10.1016/j.scitotenv.2018.10.166.

## Chapter 7. Putting bugs into politics and the economy

1. Megan E. Frederickson et al., '"Devil's Gardens" Bedevilled by Ants', *Nature* 437 (September 2005): 495–96, https://doi.org/10.1038/437495a.
2. Chloe Sorvino, 'Silent Giant: America's Biggest Private Company Reveals Its Plan to Get Even Bigger', *Forbes*, 22 October, 2018, https://www.forbes.com/sites/chloesorvino/2018/10/22/silent-giant-americas-biggest-private-company-reveals-its-plan-to-get-even-bigger-1/.
3. 'Plate Tech-Tonics: Mapping Corporate Power in Big Food,' ETC Group, 27 November 2019, https://www.etcgroup.org/content/plate-tech-tonics.
4. Sharon Anglin Treat, 'Revisiting Crisis by Design: Corporate Concentration in Agriculture', Institute for Agriculture and Trade Policy, 20 April, 2020, https://www.iatp.org/documents/revisiting-crisis-design-corporate-concentration-agriculture; 'Plate

Tech-TechTonics: Mapping Corporate Power in Big Food,' ETC Group, 27 November 2019, https://www.etcgroup.org/content/plate-tech-tonics.

5. Lisa Held, 'Bayer Forges Ahead with New Crops Resistant to 5 Herbicides', *Civil Eats*, 1 July 2020, https://civileats.com/2020/07/01/bayer-forges-ahead-with-new-crops-resistant-to-5-herbicides-glyphosate-dicamba-2-4-d-glufosinate-quizalofop/.
6. 'Mega-Mergers in the Global Agricultural Inputs Sector: Threats to Food Security & Climate Resilience', ETC Group, 30 October 2015: http://www.etcgroup.org/content/mega-mergers-global-agricultural-inputs-sector.
7. The companies: Syngenta, DuPont, Dow, Bayer and Monsanto. Cited in 'Plate Tech-Tonics: Mapping Corporate Power in Big Food', ETC Group, 2019, https://www.etcgroup.org/contentplate-tech-tonics.
8. 'Agribusiness: Lobbying, 2020', Center for Responsive Politics, accessed 5 July 2020, https://www.opensecrets.org/industries/lobbying.php?ind=A.
9. Nina Holland and Benjamin Sourice, 'Monsanto Lobbying: An Attack on Us, Our Planet and Democracy', Corporate Europe Observatory (October 2016), https://corporateeurope.org/sites/default/files/attachments/monsanto_v09_web.pdf.
10. Vicki Hird, 'Big Soy: Small Paraguayan Farmers Fighting Back against Global Agribusiness', *Guardian*, 3 February 2015, https://www.theguardian.com/global-development-professionals-network/2015/feb/03/big-soy-small-farmers-are-fighting-back-against-power-agribusiness.
11. Gregory M. Mikkelson et al., 'Economic Inequality Predicts Biodiversity Loss', *PLOS One* (May 2007), https://doi.org/10.1371/journal.pone.0000444.
12. Maike Hamann, et al., 'Inequality and the Biosphere' *Annual Review of Environment and Resources* 43 (October 2018): 61-83, https://doi.org/10.1146/annurev-environ-102017-025949.
13. 'Human Rights as an Enabling Condition in the Post-2020 Global Biodiversity Framework', International Union for Conservation of Nature, 27 February, 2020, https://www.iucn.org/news/protected-areas/202002/human-rights-enabling-condition-post-2020-global-biodiversity-framework#_ftn1.
14. Patrick Barkham, 'Ants Run Secret Farms on English Oak Trees, Photographer Discovers', *Guardian*, 24 January, 2020, https://www

.theguardian.com/environment/2020/jan/24/ants-run-secret-farms-on-english-oak-trees-photographer-discovers.

15. Thomas Wiedmann et al., ‘Scientists’ Warning on Affluence’, *Nature Communications* 11, no. 3107 (2020): https://doi.org/10.1038/s41467-020-16941-y.
16. For more on Doughnut Economics, see https://doughnuteconomics.org/about-doughnut-economics.
17. See https://betterfoodtraders.org/ for details of their aims and activities.
18. Kate Roworth, ‘Introducing the Amsterdam City Doughnut’, 8 April 2020, https://www.kateraworth.com/2020/04/08/amsterdam-city-doughnut/.

# INDEX

# ABOUT THE AUTHOR

TIM RICE

Vicki Hird is head of the Sustainable Farming Campaign for Sustain: The Alliance for Better Food and Farming, and she also runs an independent consultancy. An experienced and award-winning environmental campaigner, researcher, writer and strategist working mainly in the food, farming and environmental policy arena, Vicki has worked on government policy for many years and is the author of *Perfectly Safe to Eat? The Facts on Food.*

Vicki's passion is insects. The first pets she gave her children were a family of stick insects and she had a giraffe-necked weevil tattoo for her fiftieth birthday. Vicki also has a Masters in Pest Management and she is a Fellow of the Royal Entomological Society (FRES).

Twitter and Instagram: @vickihird